Holt Mathematics

Family Involvement Activities
Course 1

HOLT, RINEHART AND WINSTON
A Harcourt Education Company

Orlando • Austin • New York • San Diego • London

Copyright © by Holt, Rinehart and Winston

All rights reserved. No part of this publication may be reproduced or transmitted in any form or by any means, electronic or mechanical, including photocopy, recording, or any information storage and retrieval system, without permission in writing from the publisher.

Teachers using HOLT MATHEMATICS may photocopy complete pages in sufficient quantities for classroom use only and not for resale.

Printed in the United States of America

If you have received these materials as examination copies free of charge, Holt, Rinehart and Winston retains title to the materials and they may not be resold. Resale of examination copies is strictly prohibited and is illegal.

Possession of this publication in print format does not entitle users to convert this publication, or any portion of it, into electronic format.

ISBN 0-03-078239-2

CONTENTS

Chapter 1
Section A 1
Section B 5

Chapter 2
Section A 9
Section B 13

Chapter 3
Section A 17
Section B 21

Chapter 4
Section A 25
Section B 29
Section C 33

Chapter 5
Section A 37
Section B 41

Chapter 6
Section A 45
Section B 49

Chapter 7
Section A 53
Section B 57

Chapter 8
Section A 61
Section B 65
Section C 69

Chapter 9
Section A 73
Section B 77

Chapter 10
Section A 81
Section B 85

CONTENTS, CONTINUED

Chapter 11
Section A 89
Section B 93
Section C 97

Chapter 12
Section A 101
Section B 105

Family Letter
Chapter 1, Section A

What We Are Learning

Number Theory

Vocabulary
These are the math words we are learning:

base the number that is raised to an exponent

compatible numbers numbers that are close to the numbers in the problem, which may help with mental math

exponent a number that tells how many times to multiply the base by itself

exponential form a number that is written with a base and an exponent

overestimate an estimate that is greater than the exact answer

underestimate an estimate that is less than the exact answer

Dear Family,

Your child is working on ordering, comparing, and estimating with whole numbers. Your child may use place value or a number line to order or compare whole numbers. The following is an example of the way place value is used to compare whole numbers.

The National Park Service reports that they maintain and service 242,755 acres of land at the national park in Badlands, South Dakota and 241,904 acres of land at the national park in Capitol Reef, Utah. Which of these national parks has more acreage?

2 4 **2**, 7 5 5
2 4 **1**, 9 0 4
2000 > 1000

242,755 > 241,904

Start at the left and compare digits in the same place value position. Look for the first place where the values are different.

The national park in Badlands, South Dakota has more acreage than the national park in Capitol Reef, Utah.

When your child uses a number line to order numbers, make sure he or she first graphs the numbers correctly, then reads the graph from left to right. The numbers will be in order from least to greatest.

This is how your child will use a number line to order numbers.

Write the numbers in order from least to greatest.
444; 386; 742

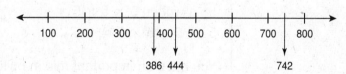

The numbers in order from least to greatest are 386, 444, and 742.

Copyright © by Holt, Rinehart and Winston.
All rights reserved.

Holt Mathematics

Family Letter
Section A, continued

Your child will be using estimation to find sums, differences, products and quotients. He or she may round to an indicated place value or use compatible numbers to estimate an answer to a problem. Here is how your child will use rounding to estimate a sum.

Estimate the sum by rounding to the place value indicated.
10,826 + 16,115; thousands

```
  11,000         Round up 10,826.
 +16,000         Round down 16,115.
  27,000
```

The sum is about 27,000.

Here is an example using compatible numbers to estimate a product.

Mike earns $5.85 an hour at the hardware store. He worked 29 hours last week. About how much money did Mike earn last week?

To find out how much money Mike earned, multiply his wage and the number of hours he worked.

$5.85 × 29 → 6 × 30   6 and 30 are compatible because they can be multiplied mentally.

6 × 30 = 180   This is an overestimate.

Mike earned about $180 last week.

Learning to find the value of numbers in exponential form is another concept that will be covered in this chapter. A number in exponential form is written with a base and an exponent. The exponent tells how many times to multiply the base by itself.

5^3 is written 5 × 5 × 5
$5^3 = 5 × 5 × 5 = 125$

You have an important role in building your child's confidence and understanding in mathematics. Your continual support will make a noticeable difference in your child's learning.

Sincerely,

Name _____ Date _____ Class _____

CHAPTER 1
Family Letter
Number Theory

Write the numbers in order from least to greatest.

1. 7,587; 6,742; 4,281; 408 _____

2. 4,212; 3,698; 1,999; 999 _____

3. 66,058; 66,842; 64,897; 6,442 _____

4. 25,103; 22,503; 23,502; 25,306 _____

Estimate each sum or difference by rounding to the place value indicated.

5. 63,765 − 32,874; ten thousands _____

6. 2,347 + 5,981; thousands _____

7. 54,879 + 89,201; ten thousands _____

8. 7,803 − 2,963; thousands _____

Write each expression in exponential form.

9. 6 × 6 _____   10. 2 × 2 × 2 × 2 × 2 _____

Find each value.

11. 3^4 _____   12. 7^3 _____   13. 4^2 _____

14. 9^1 _____   15. 5^5 _____   16. 6^3 _____

17. Bobby started with one pet rabbit. Every 6 months the number of rabbits tripled. How many rabbits were there after 2 years? Explain your answer using exponents. (*Hint:* There are 2 six-month periods every year.)

_____

Answers: 1. 408; 4,281; 6,742; 7,587 2. 999; 1,999; 3,698; 4,212 3. 6,442; 64,897; 66,058; 66,842 4. 22,503; 23,502; 25,103; 25,306 5. 30,000 6. 8,000 7. 140,000 8. 5,000 9. 6^2 10. 2^5 11. 81 12. 343 13. 16 14. 9 15. 3,125 16. 216 17. 3^4 or 81 rabbits in 2 years

Name _____ Date _____ Class _____

CHAPTER 1 Family Fun
Exponential Excitement

Materials
Gameboard
1 number cube
2–4 game pieces

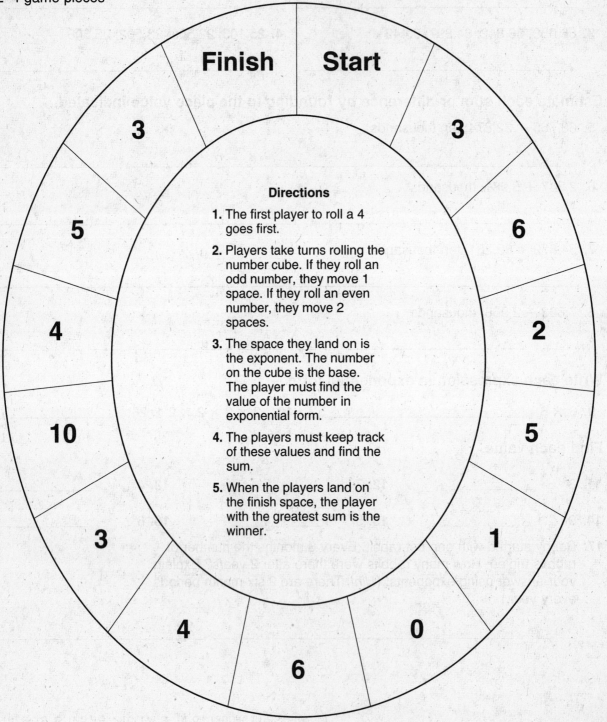

Directions

1. The first player to roll a 4 goes first.

2. Players take turns rolling the number cube. If they roll an odd number, they move 1 space. If they roll an even number, they move 2 spaces.

3. The space they land on is the exponent. The number on the cube is the base. The player must find the value of the number in exponential form.

4. The players must keep track of these values and find the sum.

5. When the players land on the finish space, the player with the greatest sum is the winner.

Family Letter
Chapter 1 Section B

What We Are Learning

Number Theory

Vocabulary
These are the math words we are learning:

arithmetic sequence a sequence in which the terms change by the same amount each time

Associative Property Changing the groupings in addend factors does not change the sum or product.

Commutative Property Changing the order of the addends or the factors does not change the sum or product.

Distributive Property Numbers added or subtracted within a set of parenthesis can be multiplied by a number outside the parenthesis.

evaluate replace each variable in an expression with a number and then follow the order of operations

numerical expression a mathematical phrase that includes only numbers and operation symbols

order of operations
1. Perform operations in parentheses.
2. Find the value of numbers with exponents.
3. Multiply or divide from left to right as ordered in the problem.

Dear Family,

Your child will need to practice following the order of operations to help her or him accurately solve many different types of problems. When there is more than one operation involved in a problem, your child will need to know which operation to do first. The steps your child needs to follow to solve multi-operational problems are listed below.

1. Perform operations in parentheses.
2. Find the value of numbers with exponents.
3. Multiply or divide from left to right as ordered in the problem.
4. Add or subtract from left to right as ordered in the problem.

Evaluate the expression $10 + 6(7 - 2)$.

$10 + 6(7 - 2)$	
$10 + 6 \times 5$	Perform the operation inside the parentheses.
$10 + 30$	Multiply.
40	Add.

Your child will be learning three number properties he or she can use to solve problems mentally.

The **Commutative Property** is an ordering property. This allows you to add or multiply numbers in any order.

addition	multiplication
$5 + 7 = 7 + 5$	$4 \times 8 = 8 \times 4$

The **Associative Property** is a grouping property. You may group the numbers in any order as long as you are only adding or only multiplying. The numbers stay in place and only the grouping symbols move. This is an example of the associative property.

addition	multiplication
$(3 + 5) + 2 = 3 + (5 + 2)$	$(5 \times 9) \times 4 = 5 \times (9 \times 4)$

The **Distributive Property** allows you to multiply a sum by a number two different ways.

$3 \times (4 + 6) = 3 \times (10) = 30$

or

$3 \times (4 + 6) = (3 \times 4) + (3 \times 6) = 12 + 18 = 30$

Family Letter
Chapter 1 Section B, continued

4. Add or subtract from left to right as ordered in the problem.

sequence an ordered set of numbers

term a number in a sequence

As your child begins to solve problems, one of the most important steps is knowing how to choose the best method of computation. Sometimes your child might use compatible numbers or one of the number properties to solve a problem using mental math. Other times, your child might use paper and pencil or a calculator.

There are many times when a calculator is a much better choice than mental math or paper and pencil. Your child will learn how to decide which method is the most appropriate choice. Be sure to ask your child **WHY** he or she chose a particular method to compute.

A strategy called *Find a Pattern* will also be introduced. Your child will learn to recognize, describe, and extend a pattern in a given sequence. One sequence that is easily recognizable is the sequence of **perfect squares.** Your child will learn that a perfect square is a whole number raised to the second power. Examine the following sequence:

1, 4, 9, 16, 25, 36, 49, 64, 81, 100, 121, 144, …

The sequence is the first 12 whole numbers raised to the second power. Ask your child to name the next term in this sequence. Your child should tell you that the next term is 13^2 or 13 raised to the second power, which equals 169.

Your child will use the four basic mathematical operations to determine the pattern of a sequence.

Identify the pattern in the sequence and name the next 3 terms.

2, 8, 7, 13, 12, 18, 17, ■, ■, ■

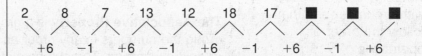

The pattern is to add 6 to the term for the next term and then subtract 1 to get the term after that. The next three terms are 23, 22, and 28.

Playing mental math games with your child will help your child exercise his or her problem solving skills. Your enthusiasm and support in your child's math activities will help to build confidence and understanding.

Sincerely,

Name _____ Date _____ Class _____

CHAPTER 1 Family Letter
Number Theory

Evaluate each expression.

1. $5 \times 6 \div 3 + (5 + 9)$
2. $8 + (9 \div 3)^2 - 10$
3. $4^2 - (16 \times 2 \div 8)^2$

Use mental math to find each sum or product.

4. $5 + 12 + 15 + 8$
5. $10 \times 2 \times 3 \times 5$
6. $49 + 64 + 11 + 26$

Use the Distributive Property to find each product.

7. 7×35
8. 9×41
9. 3×73

Which computation method would you use, mental math or a calculator, to solve the following problems?

10. For her fruit salad recipe, Suzie needs 45 apples, 21 pears, 18 plums, 19 oranges and 43 grapes. How many pieces of fruit does she need? _____

11. Out of 798 students, 302 made the honor roll. How many students did not make the honor roll? _____

Identify a pattern in each sequence and name the next three terms.

12. 4, 20, 100, 500, ■, ■, ■, …
13. 5, 16, 27, 38, ■, ■, ■, …

Identify a pattern in each sequence and name the missing terms.

14. 3, 9, ■, ■, 43,046,721, … _____

15. 1, 8, 2, ■, 4, 32, ■, 64, … _____

Answers: 1. 24 2. 7 3. 0 4. 40 5. 300 6. 150 7. $(7 \times 30) + (7 \times 5) = 210 + 35 = 245$ 8. $9 \times (40 + 1) = (9 \times 40) + (9 \times 1) = 360 + 9 = 369$ 9. $3 \times (70 + 3) = (3 \times 70) + (3 \times 3) = 210 + 9 = 219$ 10. calculator 11. mental math 12. multiply each term by 5 to get the next term: 2,500, 12,500, 62,500 13. add 11 to each term to get the next term: 49, 60, and 71 14. square each term; 81 and 6,561 15. multiply one term by eight and divide the next term by four: 16 and 8

Name _____ Date _____ Class _____

Family Fun
Operation Riddle

Solve each expression. Match the answer with the letter code below to find the answer to the following question.

Question: What is a cool place in which people compete?

Answer: _____

E	S	M	W	L	Y	N	T	R	P	C	I	O
2	4	68	25	0	28	9	96	27	16	1	5	300

Solve each expression.	Answer	Letter
$(2 + 4) \times 5 - 14 + 9$		
$8 \times 3 \div 2 - (21 \div 3)$		
$6 + 2 + 3 \times 3 - 8$		
$(7 - 1) + 6 \times (2 + 13)$		
$3 \times 4 - 8 + 5 - 7$		
$(20 + 7) \div 9 \times (13 - 4)$		
$5^2 \times (47 + 13) \div 5$		
$4 \times 4 \times 4 - (12 \div 3)^3$		
$30 - 8 \div 4$		
$(10^2 - 2^5)$		
$19 + 21 - 14 - 5 \times 2$		
$10 - (2 + 3)$		
$(8 + 17) \div (2 + 3)^2$		
$48 \div (4^2 - 4)$		

Answers: Winter Olympics

Family Letter
Chapter 2, Section A

What We Are Learning

Understanding Variables and Expressions

Vocabulary
These are the math words we are learning:

algebraic expression an expression that contains one or more variables and may contain operation symbols

constant a quantity that does not change

variable a letter or symbol that represents a quantity that can change

Dear Family,

Your child is learning how to identify and evaluate algebraic expressions. When evaluating algebraic expressions, your child will substitute a value for the variable and then simplify the expression. An example of how to do this is below.

Evaluate each expression to find the missing values in the table.

w	w • 15
10	150
12	■
15	■

Substitute the values for w in $w \cdot 15$.
$w = 10$; $10 \cdot 15 = 150$
$w = 12$; $12 \cdot 15 = 180$
$w = 15$; $15 \cdot 15 = 225$

The missing values are 180 and 225.

Your child will also learn how to use a table to find an algebraic expression. This is how your child will find an expression.

Find an expression for the table.

x	■
48	6
40	5
32	4

What pattern do you see?
$48 \div 8 = 6$
$40 \div 8 = 5$
$32 \div 8 = 4$

Each value for x is being divided by 8, so the expression for this table is $x \div 8$.

Whether your child is finding the value for a variable or finding the algebraic expression for a table, he or she is on the way to building a solid base for comprehending algebraic concepts.

A continuing step in solving algebraic problems is to know what action or operation to perform. Your child will learn the words that suggest the correct mathematical operation. You can help reinforce this concept by giving your child various scenarios and having her or him translate the words into math expressions and math expressions into words.

Family Letter

Section A, continued

The following is a chart students will use to help decipher the math terminology.

Action/Words	Operation	Examples
Put together, sum, plus, added to, more than	Addition	32 plus 5 → 32 + 5 $x + 15$ → 15 added to x
Subtracted from, minus, take away, the difference of, less than	Subtraction	$s - 12$ → 12 less than s $12 - 5$ → 12 minus 5 $5 - m$ → m subtracted from 5
Times, multiplied, product, doubling, tripling	Multiplication	$12a$ → 12 times a $3 \cdot 6$ → the product of 3 and 6
Divided by, quotient	Division	$125 \div 25$ → the quotient of 125 and 5 $n \div 3$ → n divided by 3

This is how your child will translate words into math expressions and math expressions into words.

Write each phrase as a numerical or algebraic expression.

A. the product of x and 14   **14x**

B. the difference of 127 and 115   **127 − 115**

Write two phrases for the expression 45x.

the product of 45 and x 45 times x

Reviewing these math terms with your child will help reinforce his or her math vocabulary and build your child's confidence in understanding variables and expressions.

Sincerely,

Name _____ Date _____ Class _____

CHAPTER 2 Family Letter
Understanding Variables and Expressions

Evaluate each expression to find the missing values in the tables.

1.
a	$a + 12$
52	64
67	
89	

2.
x	$5x$
11	55
12	
13	

Find an expression for each table.

3.
w	▮
18	12
28	22
38	32

4.
r	▮
12	6
14	7
16	8

5. Is $3x + 15$ an algebraic expression? _____

6. Evaluate $x + 20$ for $x = 7$. _____

7. Evaluate $12s$ for $s = 3$. _____

Write an expression.

8. A watch cost d dollars. How would you write the cost of a radio that costs $15 more than the watch?

_____

9. If you divide n newspapers into 5 equal piles, how many newspapers are in each pile?

_____

Write each phrase as a numerical or algebraic expression.

10. 35 less than p

_____

11. the sum of 52 and 108

_____

12. Write two phrases for the expression $d + 23$.

_____

Answers: 1. 79; 101 2. 60; 65 3. $w - 6$ 4. $r \div 2$ 5. Yes 6. 27 7. 36 8. $d + 15$ 9. $n \div 5$ 10. $p - 35$ 11. $52 + 108$ 12. d plus 23; 23 more than d; 23 added to d; the sum of d and 23

Holt Mathematics

Name _____ Date _____ Class _____

CHAPTER 2

Family Fun
Mind-Reading Mathematics

If you have ever wanted to impress your friends and family by being able to read their minds, now is your chance!

Just follow these simple directions for two amazing tricks to become a Mind-Reading Mathematician.

Game 1
I Can Guess Your Answer

1. Think of a number.

2. Double it.

3. Add 6.

4. Divide it by 2.

5. Subtract your original number from the quotient.

What is your answer every single time?
_____

Game 2
I Know What You're Thinking

1. Think of a number between 1 and 9.

2. Multiply that number by 7.

3. Multiply that product by 3.

4. Multiply that product by 11.

5. Multiply that product by 37.

6. Multiply that product by 13.

What is your answer every single time?
_____

Answers: Game 1. 3 Game 2. Six digit number with the same digit repeated. Sample: 111,111

Family Letter
Chapter 2, Section B

What We Are Learning

Equations

Vocabulary

These are the math words we are learning:

equation a mathematical statement that says two quantities are equal

solution a value that makes the equation true

Dear Family,

In the previous lesson, your child learned to translate words into numbers, variables, and operations. Your child will build on this by learning to solve equations. The first step in solving equations is understanding that the solution for the equation makes the equation true. Your child will learn how to determine if a specific value is a solution of an equation.

This is how your child will determine if a given value is a solution to an equation.

Determine whether the given value of the variable is a solution.

A. $b - 9 = 22$ for $b = 27$

$27 - 9 \stackrel{?}{=} 22$ Substitute 27 for b.

$18 \stackrel{?}{=} 22$ Subtract.

$18 \neq 22$

Since 18 does not equal 22, 27 is not a solution to $b - 9 = 22$.

B. $17v = 102$ for $v = 6$

$17 \cdot 6 \stackrel{?}{=} 102$ Substitute 6 for v.

$102 \stackrel{?}{=} 102$ Multiply.

Since $102 = 102$, then 6 is a solution to $17v = 102$.

Your child will use this skill to check the solutions to the equations he or she will be solving.

Once your child knows how to check if a solution to an equation is true, he or she will learn how to solve four different types of whole number equations. These equations involve addition, subtraction, multiplication, and division. For each equation type, your child will learn how to "undo" the given operation by performing the inverse or opposite operation.

One of the most important and often overlooked steps in equation solving is checking to make sure the solution is correct. Reinforce with your child the importance of checking his or her solution.

Family Letter

Chapter 2 Section B, continued

Solve each equation. Check your answers.

Addition

$x + 77 = 115$
$\underline{-77 = -77}$
$x = 38$

77 is added to x.
Subtract 77 from both sides to undo the addition.

Check: $x + 77 \stackrel{?}{=} 115$
$38 + 77 \stackrel{?}{=} 115$
$115 = 115$ ✓

Substitute 38 for x in the equation.
38 is the solution.

Subtraction

$k - 14 = 35$
$\underline{+14 \quad +14}$
$k = 49$

14 is subtracted from k.
Add 14 to both sides to undo the subtraction.

Check: $k - 14 \stackrel{?}{=} 35$
$49 - 14 \stackrel{?}{=} 35$
$35 = 35$ ✓

Substitute 49 for k in the equation.
49 is the solution.

Multiplication

$15t = 75$
$\dfrac{15t}{15} = \dfrac{75}{15}$
$t = 5$

t is multiplied by 15.
Divide both sides by 15 to undo the multiplication.

Check: $15t \stackrel{?}{=} 75$
$15(5) \stackrel{?}{=} 75$
$75 = 75$ ✓

Substitute 5 for t in the equation.
5 is the solution.

Division

$\dfrac{a}{12} = 8$
$12 \cdot \dfrac{a}{12} = 8 \cdot 12$
$a = 96$

a is divided by 12.
Multiply both sides by 12 to undo the division.

Check: $\dfrac{a}{12} \stackrel{?}{=} 8$
$\dfrac{96}{12} \stackrel{?}{=} 8$
$8 = 8$ ✓

Substitute 96 for a in the equation.
96 is the solution.

Solving for x can be challenging and exciting. Encourage your child to practice making up different whole number equations and challenge each other to see who solves them more quickly.

Sincerely,

Name _____ Date _____ Class _____

CHAPTER 2 Family Letter
Equations

Determine whether the given value of the variable is a solution.

1. $3 + h = 71$ for $h = 68$

2. $g - 6 = 15$ for $g = 10$

3. $\frac{r}{10} = 9$ for $r = 90$

4. $16t = 84$ for $t = 4$

Solve each equation. Check your answers.

5. $x + 36 = 48$

6. $\frac{p}{8} = 12$

7. $u - 17 = 32$

8. $36n = 36$

9. $44 = w + 38$

10. $\frac{q}{16} = 5$

Write and solve an equation.

11. Carole spent $45 on shoes. After her purchase, she had $58 left. How much money did she start with?

12. Brian has 315 stamps in his collection. One stamp book holds 35 stamps. How many books does Brian need to hold his whole collection?

Answers: 1. yes 2. no 3. yes 4. no 5. $x = 12$ 6. $p = 96$ 7. $u = 49$ 8. $n = 1$ 9. $w = 6$ 10. $q = 80$ 11. $x - 45 = 58$; She started with $103. 12. $315 = 35s$; He needs 9 books.

Holt Mathematics

Name _____ Date _____ Class _____

CHAPTER 2 Family Fun
SEA-MAIL

Solve each equation. Then put the letter of the variable above its value to answer the riddle.

$14d = 84$ d = _____

$32 = y + 21$ y = _____

$\dfrac{o}{4} = 6$ o = _____

$r - 3 = 15$ r = _____

$25c = 125$ c = _____

$s + 17 = 26$ s = _____

$\dfrac{n}{0.5} = 8$ n = _____

$b - 7 = 5$ b = _____

$180 = 12e$ e = _____

How did the Vikings send secret messages?

___ ___   ___ ___ ___ ___ ___   ___ ___ ___ ___
12   11   4   24   18   9   15   5   24   6   15

Answer: By Norse Code

Family Letter

Chapter 3, Section A

What We Are Learning

Understanding Decimals

Vocabulary
These are the math words we are learning:

clustering rounding all of the numbers to the same value

front-end estimation estimating with only the whole number part of the decimal

Dear Family,

Now that your child has an understanding of whole number applications, he or she can begin to apply that knowledge to decimals. Examples of how your child will model decimals, read and write decimals, and compare and order decimals are shown below.

Model each decimal.

0.17   1.48

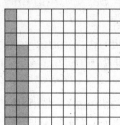  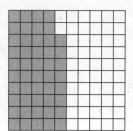

Write each decimal in expanded form, standard form, and words.

Standard Form	Expanded Form	Words
3.06	3 + 0.06	Three and six hundredths
51.128	51 + 0.128	Fifty-one and one hundred twenty-eight thousandths

Ask your child what the word *"and"* stands for when reading and writing decimals. Your child should tell you that the word *"and"* stands for the decimal point.

Your child can use place value or a number line to compare and order decimals. These are the same methods your child used to compare and order whole numbers.

Compare 21.73 and 21.77.

21.73
21.77   **Step 1** Line up the decimal points.

21.73
21.77   **Step 2** Start at the left and compare digits.

2 1.7 **3**
2 1.7 **7**   **Step 3** Look for the first place where the digits are different.

Since 7 > 3, then 21.77 > 21.73.

Holt Mathematics

CHAPTER 3

Family Letter

Section A, continued

Your child will be using estimation to find sums, differences, products, and quotients. Students may round to an indicated place value or use compatible numbers to estimate an answer to a problem. Two new estimating techniques, clustering and front-end estimation will be introduced to your child in this section. The following is an example of front-end estimation.

Estimate the sum using front-end estimation.
14.75 + 32.22 + 6.19

14 + 32 + 6 = 52   Add only the whole numbers.

Because the whole number values of the decimals are less than the actual numbers, the sum is an *underestimate*. Therefore the exact answer is 52 or more.

The addition and subtraction of decimals is very similar to that of whole numbers. Your child will learn how to accurately add and subtract decimals by knowing where to correctly place the decimal point and when to add place-holding zeros. For example:

Find the difference. 12.3 − 7.56

12.30	Align the decimal points.
− 7.56	Use zero as a placeholder.
4.74	Subtract. Place the decimal point.

Your child will learn how to multiply and divide decimals by powers of ten just by moving the decimal point. This concept will lead your child directly to converting metric measurements.

Multiply or divide.

34.21×10^4 The power of ten is 4. To multiply, move the decimal point 4 places to the right.
= 342,100   Add two zeros as placeholders.

$34.21 \div 10^4$ The power of ten is 4. To divide, move the decimal point 4 places to the left.
= 0.003421   Add two zeros as placeholders.

Continue to review these decimal concepts daily with your child.

Sincerely,

Name _____ Date _____ Class _____

CHAPTER 3 Family Letter
Understanding Decimals

Model each decimal.

1. 0.3   **2.** 0.76   **3.** 1.51

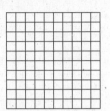   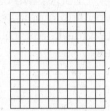   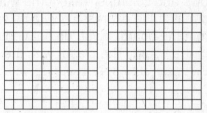

Write each number in standard form, expanded form, and words.

4. 3.103   **5.** 41 + 0.63   **6.** 1 and eight-tenths

_____   _____   _____

_____   _____   _____

Order the decimals from least to greatest.

7. 25.12, 25.07, 25.5   **8.** 7.33, 7.35, 7.3

_____   _____

Estimate. Round to the indicated place value.

9. 3.0567 + 7.123; hundredths   **10.** 95.63 − 74.09; tenths

_____   _____

Find the sum or difference.

11. 42.18   **12.** 5.03   **13.** 18   **14.** 39.12   **15.** 8.3
   + 0.05   − 0.15   − 1.93   + 1.3   − 1.2

Multiply or divide.

16. 7.8 × 10,000   **17.** 1.5 ÷ 10^3 **18.** 42,516 ÷ 100   **19.** 0.25 × 10^5

_____   _____   _____   _____

Convert each measure.

20. 2.4 m = _____ mm   **21.** 350 g = _____ kg

Answers: 1. 2. 3. 4. 3 + 0.1 + 0.003; three and one hundred three thousandths 5. 41.63; forty-one and sixty-three hundredths 6. 1.8; 1 + 0.8 7. 25.07, 25.12, 25.5 8. 7.3, 7.33, 7.35 9. 10.18 10. 21.5 11. 42.23 12. 4.88 13. 16.07 14. 40.42 15. 7.1 16. 78,000 17. 0.0015 18. 425.16 19. 25,000 20. 2,400 mm 21. 0.35 kg

19   Holt Mathematics

Family Fun
Chapter 3: Decimal Concentration

Directions
Cut out the cards below. Shuffle the cards and place them face down in rows and columns. Take turns with your partner. Choose two cards and try to match the expression with its value. When you find a pair, remove it from the board. If you do not find a match, return the cards to their spots on the board. The player with the most matches wins.

6.02	six and 2 hundredths	850 g	0.85 kg	Numbers listed from greatest to least
9.602 9.6 9.062 9.006	2.089	2 + 0.089	78.9 −51.06	About 28
8.97×10^6	8,970,000	1.06 0.98 + 10.99	about 13	0.4 ÷ 100
0.004	73 cm	730 mm	height of a man	About 1.5 m

Family Letter
Chapter 3, Section B

What We Are Learning

Multiplying and Dividing Decimals

Vocabulary
These are the math words we are learning:

scientific notation
a shorthand method for writing large numbers

Dear Family,

In this section, your child will continue to learn about decimals and decimal applications. One such application involves using decimals to express very large numbers in a simple way. This is called scientific notation. An example is shown below.

Write 6,358,000 in scientific notation.

Step 1   Move the decimal point left to form a number greater than one but less than ten.

6.358000

$6.358000 > 1$
$6.358000 < 10$

Step 2   Since the decimal point was moved 6 places to the left, the number should be multiplied by ten to the sixth power. 10^6

Step 3   Write the number in scientific notation.

$6,358,000 = 6.358 \times 10^6$

Your child will also learn how to write a number in standard form when given a number written in scientific notation.

Write 7.82×10^5 in standard form.

7.82×10^5
7.82000   The power of 10 is 5. Rewrite the number with zeros.
782,000   Now, move the decimal point 5 places to the right.

So, $7.82 \times 10^5 = 782,000$.

Until now, your child has learned to multiply and divide decimals by the power of ten. In this section, your child will learn to multiply a decimal by a whole number and by another decimal. He or she will also learn to divide a decimal by a whole number and by a decimal.

Becoming proficient with these types of problems will allow your child to evaluate products and quotients for given variables, and learn to solve decimal equations.

The following shows examples of the steps your child will use to multiply and divide decimals. Notice that the process is similar to the one used to multiply and divide whole numbers except that you have to place a decimal point in the product or quotient.

Family Letter

Chapter 3 Section B, continued

Multiply a Decimal by a Whole Number

6 × 0.4

6 × 0.4 → Multiply as you would with whole numbers.

Add the number of decimal places in each number multiplied.

6 → 0 decimal place
0.4 → + 1 decimal place
       1 decimal place

6 × 0.4 = 2.4 → Place the decimal point **1** digit to the left.

Multiply a Decimal by a Decimal

0.6 × 0.4

0.6 × 0.4 → Multiply as you would with whole numbers.

Add the number of decimal places in each number multiplied.

0.6 → 1 decimal place
0.4 → + 1 decimal place
       2 decimal places

0.6 × 0.4 = 0.24 → Place the decimal point **2** digits to the left.

Divide a Decimal by a Whole Number

15.81 ÷ 3

Place the decimal point in the quotient directly above the decimal point in the dividend.

```
     5.27
  3)15.81    Divide as you
   -15       would with
     8       whole
    -6       numbers.
    21
   -21
     0
```

Divide a Decimal by a Decimal

15.81 ÷ 0.03

Make the divisor, 0.03, a whole number by multiplying it and the dividend, 15.81, by the same power of ten.

0.03 × 100 = 3
15.81 × 100 = 1,581

```
     527
  3)1581    Divide
   -15
     8
    -6
    21
   -21
     0
```

It is important that your child understand the reasonableness of an answer when performing decimal operations. When solving equations with decimal coefficients, have your child explain the answers to make certain that the solutions are reasonable.

Sincerely,

Name _____ Date _____ Class _____

CHAPTER 3 Family Letter
Multiplying and Dividing Decimals

Write each number in scientific notation.

1. 67,003 _____   2. 974,875 _____   3. 1,000,000 _____

Write each number in standard form.

4. 3.8×10^4 _____   5. 2.8×10^2 _____   6. 7.82×10^6 _____

Find each product.

7. 0.3
   × 0.6

8. 45
   × 0.4

9. 2.71
   × 8.0

Write and solve an equation.

10. A sheet of drywall is on sale for $2.89. How much will 25 sheets of drywall cost?

_____

Evaluate 13x for each value of x.

11. $x = 0.35$ 12. $x = 1.51$ 13. $x = 7.8$

Evaluate $x \div 8$ for each value of x.

14. $x = 0.824$ 15. $x = 40.6$ 16. $x = 16.016$

Find each quotient.

17. $7.2 \div 1.2$ 18. $71.4 \div 0.03$ 19. $68.76 \div 1.8$

Solve.

20. The soccer team raised $260.75 for new uniforms. Each uniform costs $26. How many new uniforms can the soccer team purchase?

21. There are 22 students in Mrs. Field's classroom. If juice boxes come in packs of 4, how many packs does Mrs. Field need for her class?

Answers: 1. 6.7003×10^4 2. 9.74875×10^5 3. 1×10^6 4. 38,000 5. 280 6. 7,820,000 7. 0.18 8. 18 9. 21.68 10. $72.25 11. 4.55 12. 19.63 13. 101.4 14. 0.103 15. 5.075 16. 2.002 17. 6 18. 2,380 19. 38.2 20. 10 uniforms 21. 6 packs

23 Holt Mathematics

Name _____ Date _____ Class _____

CHAPTER 3

Family Fun
Capture the Flag

Directions
Solve each problem. Then find the sum of the digits of your answer. Shade the square if the sum is an odd number. The picture made by the shaded squares will tell you under which block the flag is. If you evaluate each number correctly, you will capture your enemy's flag.

22×2.6	0.67×10^3	$8.7 \div 0.3$	12×10^2
20×10^5	18.2×5.1	0.5×0.4	3.21×7
$10.55 \div 5$	$9.2 \div 0.5$	6.81×3	$27.6 \div 100$
1.5×0.5	$19 \div 0.2$	33×100	$98 \div 0.7$
9.8×0.4	$1.1 \div 0.5$	$18.3 \div 6$	0.6×0.3

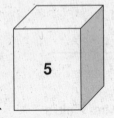
5

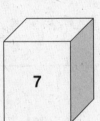
7

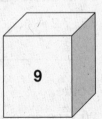
9

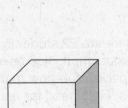
1

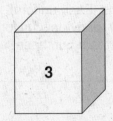
3

Answer: The flag is under the 9 block.

Family Letter
Section A

What We Are Learning

Number Theory

Vocabulary
These are the math words we are learning:

composite number numbers that are divisible by more than two numbers

divisible able to be divided by a number without having a remainder

factor whole numbers that are multiplied to find a product

greatest common factor the largest of the common factors shared by two or more whole numbers

prime factorization the process of writing the number as the product of its prime factors

prime number a number divisible by only the number 1 and itself

Dear Family,

Your child is working with divisibility rules, finding prime factors, and determining the greatest common factor of whole numbers. These skills are precursors to the more advanced applications involving fractions.

Knowledge of divisibility rules will allow your child to use mental math and number patterns to determine if a number can be evenly divided into another number. Knowing these rules will also allow your child to see the relationship and patterns in numbers. Have your child use the guidelines and the examples in the table to explain how one determines the divisibility of a particular number.

Divisibility Rules

A number is divisible by ...	Divisible	Not Divisible
2 if the last digit is an even number.	7,992	4,221
3 if the sum of the digits is divisible by 3.	219	451
4 if the last two digits form a number divisible by 4.	5,516	431
5 if the last digit is 0 or 5.	795	977
6 if the number is divisible by both 2 and 3.	132	561
9 if the sum of the digits is divisible by 9.	801	1,712
10 if the last digit is 0.	27,000	1,895

Tell whether the number 846 is divisible by 2, 3, 4, 5, and 6.

2	The last digit is an even number.	Divisible
3	The sum of the digits is 18, which is divisible by 3.	Divisible
4	The last two digits form the number 46. 46 is not divisible by 4.	Not divisible
5	The last digit is not a 0 or a 5.	Not divisible
6	The number is divisible by both 2 and 3.	Divisible

846 is divisible by 2, 3, and 6.

Family Letter
Chapter 4 — Section A, continued

Divisibility rules are the tricks of the trade. However, you need to remember that all numbers are divisible by their factors. Some numbers are divisible by many factors. These are called **composite numbers**.

A number that is only divisible by the number 1 and itself is called a **prime number**. Your child will be writing the factors of a number in terms of the product of that number's prime factors. This is called **prime factorization**.

Write the prime factorization of the number 84.
Method 1: Use a tree diagram.

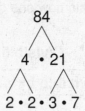

Choose any two factors of 84. Keep finding factors until each branch ends with a prime factor.

The prime factorization of 84 is 2 • 2 • 3 • 7 or 2^2 • 3 • 7.

Method 2: Use a ladder diagram.

```
 |84
2|42
  2|21
    3|7
     7|1
```

Choose a prime factor of 84 to begin. Keep dividing by prime factors until the quotient is 1.

Again, the prime factorization of 84 is 2 • 2 • 3 • 7 or 2^2 • 3 • 7. Notice how both methods result in the same answer.

Your child will use prime factorization as one way to help find the greatest common factor, GCF, of a set of numbers. Your child will list the prime factors of each number in the set, circle the matching factors, and then multiply those common prime factors. Knowing how to find the GCF is essential when working with fractions.

This section contains important skills that will allow your child to comprehend the concepts and applications that will be studied throughout the rest of this chapter. Review these skills often to help your child become more confident in this topic.

Sincerely,

Name _____ Date _____ Class _____

Chapter 4 Family Letter
Number Theory

Tell whether each number is divisible by 2, 3, 4, 5, 6, 9, or 10.

1. 342
2. 981
3. 200
4. 836

Replace each question mark with a digit that will make the number divisible by 6.

5. 47?
6. 8?0
7. ?28
8. 2?

Tell whether each number is prime or composite.

9. 61
10. 93
11. 39
12. 19

List all the factors of each number.

13. 42
14. 88
15. 24
16. 50

Write the prime factorization of each number.

17. 64
18. 18
19. 55
20. 48

Find the GCF of each set of numbers.

21. 12 and 50
22. 15 and 75
23. 18, 24, and 30

24. Brenda is making toy baskets to donate to charity. She has 24 toy cars, 30 jump ropes, and 42 teddy bears. What is the greatest number of baskets she can make if each type of toy is equally distributed among the baskets?

Answers: 1. Yes for 2, 3, 6, 9 2. Yes for 3, 9 3. Yes for 2, 4, 5, 10 4. 1, 4, or 7 7. 2, 5, or 8 8. 4. 9. Prime 10. Prime 11. Composite 12. Composite 13. 1, 2, 3, 6, 7, 14, 21, 42 14. 1, 2, 4, 8, 11, 22, 44, 88 15. 1, 2, 3, 4, 6, 8, 12, 24 16. 1, 2, 5, 10, 25, 50 17. 2^6 18. $2 \cdot 3^2$ 19. 5 · 11 20. $2^4 \cdot 3$ 21. 2 22. 15 23. 6 24. 6 baskets

Name _____ Date _____ Class _____

CHAPTER 4
Family Fun
Divisibility Squares

Directions

- If a number is divisible by 5, color it blue.
- If a number is divisible by 6, keep it white.
- If a number is divisible by 2 but not by 5 or 6, color it red.

154	32	44	254	392	76
324	912	36	822	762	712
442	18	55	975	108	916
184	444	305	415	234	886
62	12	336	888	126	16
746	98	202	694	242	512

Create your own divisibility square design and pass it on to a friend.

Answers: Blue: 55, 975, 305, 415; White: 324, 912, 36, 822, 762, 18, 108, 444, 234, 12, 336, 888, 126; Red: 154, 32, 44, 254, 392, 76, 712, 442, 916, 184, 886, 62, 16, 746, 98, 202, 694, 242, 512

Family Letter
Chapter 4, Section B

What We Are Learning

Understanding Fractions

Vocabulary
These are the math words we are learning:

equivalent fractions fractions that represent the same value

improper fraction a fraction whose denominator is larger than its numerator; its value is greater than 1

mixed number a number that contains both a whole number greater than zero and a fraction

proper fraction a fraction whose denominator is smaller than its numerator; its value is less than 1

repeating decimal a decimal with a block of one or more digits that repeat without end

simplest form a fraction where the greatest common factor of the numerator and the denominator is 1

terminating decimal a decimal with a finite number of decimal places

---

Dear Family,

Your child is beginning to study fractions. Some of the basic skills your child will learn in this section are converting between decimals and fractions and writing equivalent fractions.

Decimals are another way to write fractions and mixed numbers. **Mixed numbers** are special fractions that consist of a whole number and a fraction. Your child will use place value concepts to convert between decimals and fractions.

Write each decimal as a fraction or mixed number.

A. 0.17

0.17 — Identify the place value of the digit farthest to the right.

$\frac{17}{100}$ — The 7 is in the hundredths place, so use 100 as the denominator.

B. 2.3

2.3 — Identify the place value of the digit farthest to the right.

$2\frac{3}{10}$ — Write the whole number, 2. The 3 is in the tenths place, so use 10 as the denominator.

When converting from a fraction to a decimal, you may discover that the decimal part will repeat itself. This type of decimal is called a **repeating decimal**. A decimal that has a finite number of decimal places is called a **terminating decimal**.

repeating decimal
$0.35\overline{35}$

terminating decimal
0.64

To convert from a fraction to a decimal, your child will divide the numerator by the denominator. The quotient will be the decimal.

Change $\frac{3}{5}$ to a decimal.

$\frac{3}{5} = 5\overline{)3.0}$ — Divide the numerator by the denominator.
  $\underline{-3.0}$
  0

$\frac{3}{5} = 0.6$ — The quotient is the decimal.

Family Letter

Chapter 4 — Section B, continued

Understanding equivalent fractions will help your child have a better grasp of how fractions are related to each other. To find a fraction equal to another fraction, simply multiply or divide both the numerator and the denominator of the given fraction by the same number.

Write two equivalent fractions for $\frac{3}{8}$.

$\frac{3 \times 2}{8 \times 2} = \frac{6}{16}$ Multiply the numerator and the denominator by 2.

$\frac{3 \times 5}{8 \times 5} = \frac{15}{40}$ Multiply the numerator and the denominator by 5.

So $\frac{3}{8}$, $\frac{6}{16}$, and $\frac{15}{40}$ are all equivalent fractions.

Ask your child to name two more fractions equivalent to $\frac{3}{8}$.

Fractions with different denominators are called **unlike fractions.** When asked to compare unlike fractions, your child will need to use the least common denominator (LCD) to create equivalent fractions. Once the fractions have common denominators, they are called **like fractions** and your child will be able to order and compare the numerators.

Compare. Write $<$, $>$, or $=$.

$\frac{2}{3} \square \frac{3}{4}$

$\frac{2}{3} = \frac{?}{12}$ $\frac{3}{4} = \frac{?}{12}$ Find a common denominator by multiplying the denominators $3 \cdot 4 = 12$.

$\frac{2 \cdot 4}{3 \cdot 4} = \frac{8}{12}$ $\frac{3 \cdot 3}{4 \cdot 3} = \frac{9}{12}$ Find equivalent fractions with 12 as the denominator.

$\frac{2}{3} = \frac{8}{12}$ $\frac{3}{4} = \frac{9}{12}$

Compare the like fractions.

$\frac{8}{12} < \frac{9}{12}$ so $\frac{2}{3} < \frac{3}{4}$.

Your child will continue to learn about fractions in the following sections.

Sincerely,

Name _____ Date _____ Class _____

CHAPTER 4 Family Letter
Understanding Fractions

Write each decimal as a fraction or a mixed number.

1. 0.37 2. 0.09 3. 6.11 4. 1.2

Write each fraction or mixed number as a decimal.

5. $3\frac{2}{5}$ 6. $\frac{3}{8}$ 7. $4\frac{1}{9}$ 8. $\frac{7}{12}$

Order the fractions and decimals from least to greatest.

9. $0.38, \frac{1}{3}, \frac{3}{10}$ 10. $\frac{8}{15}, \frac{1}{2}, 0.75$

Find two equivalent fractions for each given fraction.

11. $\frac{12}{16}$ 12. $\frac{11}{22}$ 13. $\frac{5}{9}$ 14. $\frac{14}{21}$

Find the missing number that makes the fractions equivalent.

15. $\frac{4}{5} = \frac{?}{20}$ 16. $\frac{9}{12} = \frac{3}{?}$ 17. $\frac{9}{10} = \frac{36}{?}$

Write each fraction in simplest form.

18. $\frac{6}{18}$ 19. $\frac{12}{15}$ 20. $\frac{25}{40}$

Compare. Write <, >, or =.

21. $\frac{3}{5} \square \frac{4}{5}$ 22. $\frac{11}{15} \square \frac{2}{3}$ 23. $\frac{9}{33} \square \frac{3}{11}$

24. Harry needs $3\frac{3}{8}$ feet of wood to make a birdhouse. Write $3\frac{3}{8}$ as an improper fraction. _____

Answers: 1. $\frac{37}{100}$ 2. $\frac{9}{100}$ 3. $6\frac{11}{100}$ 4. $1\frac{2}{10}$ 5. 3.4 6. 0.375 7. $4.1\overline{1}$ 8. $0.58\overline{3}$ 9. $0.38, \frac{3}{10}, \frac{1}{3}$ 10. $\frac{8}{15}, \frac{1}{2}, 0.75$ Possible Answers for 11–14: 11. $\frac{24}{32}, \frac{3}{4}$ 12. $\frac{1}{2}, \frac{33}{66}$ 13. $\frac{10}{18}, \frac{15}{27}$ 14. $\frac{2}{3}, \frac{28}{42}$ 15. 16 16. 4 17. 40 18. $\frac{1}{3}$ 19. $\frac{4}{5}$ 20. $\frac{5}{8}$ 21. < 22. > 23. = 24. $\frac{27}{8}$

Name _____ Date _____ Class _____

CHAPTER 4

Family Fun
First Team Home Wins!

Materials

1 number cube
2 markers (coins, paper clips, etc)

Directions

- Roll the number cube to see which team goes first. The lowest roll goes first. Each team takes turns.

- Team 1 rolls the cube and moves that many places on the game board.

- Team 1 then has to name a fraction equivalent to the fraction in the space where they landed.

- Team 2 checks to make sure the two fractions are equivalent. If Team 1 is correct, they may stay on that space. If Team 1 is incorrect, they must roll the number cube and move back the number of spaces it shows.

- The first team to reach the finish line wins!

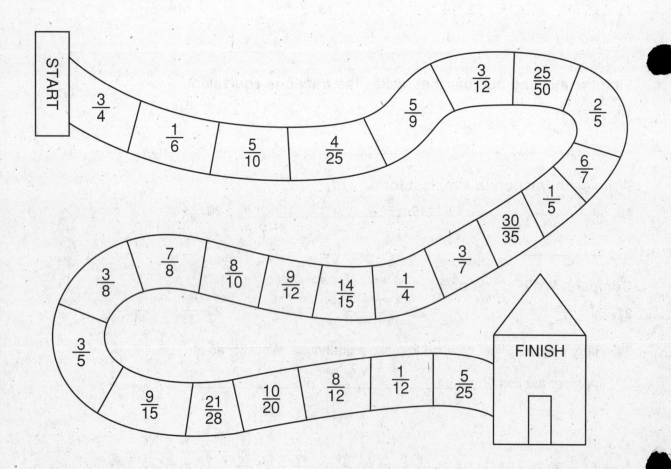

Copyright © by Holt, Rinehart and Winston.
All rights reserved.

Holt Mathematics

Family Letter
Chapter 4 Section C

What We Are Learning

Introduction to Fraction Operations

Dear Family,

In this section, your child will use the concepts and skills from previous sections to add and subtract like fractions and mixed numbers. Once the sum or difference has been calculated, your child will need to remember to write the answer in simplest form.

Add. Write your answer in simplest form.

$\frac{2}{9} + \frac{4}{9}$

$\frac{2}{9} + \frac{4}{9} = \frac{6}{9}$ Add the numerators. Keep the same denominator.

$= \frac{2}{3}$ Write your answer in simplest form.

Remember, when the numerator equals the denominator, the fraction is equal to 1. For instance: $\frac{10}{10}$ and $\frac{5}{5}$ both equal 1.

Subtract. Write your answer in simplest form.

$1 - \frac{5}{8}$

$\frac{8}{8} - \frac{5}{8}$ To get a common denominator, rewrite 1 as a fraction with the denominator of 8.

$\frac{8}{8} - \frac{5}{8} = \frac{3}{8}$ Subtract the numerators. Keep the denominator the same.

Your child will also learn to evaluate expressions by substituting a given value into an expression and then performing the indicated operation, addition or subtraction. This skill will help prepare your child for the more complex fraction applications that will appear in the upcoming chapters, as well as other math courses.

Family Letter
Section C, continued

When fractions have different denominators, you can use estimation to find sums and differences. Rounding fractions to 0, $\frac{1}{2}$, or 1, can help you figure out what numbers to add or subtract. Looking at the number line will help you decide what number the fraction should be rounded to.

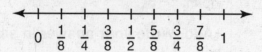

$\frac{1}{8}$ should be rounded to 0 because it is closest to 0. $\frac{3}{8}$ rounds to $\frac{1}{2}$ and $\frac{3}{4}$ rounds to 1 because they are closest to 1. You can then use these rounded fractions to add and subtract.

Find the sum.

$\frac{2}{12} + \frac{15}{16}$ $\frac{2}{12}$ rounds to 0. $\frac{15}{16}$ rounds to 1.

$0 + 1 = 1$

$\frac{2}{12} + \frac{15}{16}$ is about equal to 1.

Find the difference.

$\frac{18}{20} - \frac{4}{10}$ $\frac{18}{20}$ rounds to 1. $\frac{4}{10}$ rounds to $\frac{1}{2}$.

$1 - \frac{1}{2} = \frac{1}{2}$

$\frac{18}{20} - \frac{4}{10}$ is about equal to $\frac{1}{2}$.

Adding and subtracting fractions is an important skill that will be useful to your child in future chapters. Provide problems for your child to practice on to strengthen his or her skills.

Sincerely,

CHAPTER 4 Family Letter
Introduction to Fraction Operations

Write each sum or difference in simplest form.

1. $\dfrac{5}{17} + \dfrac{10}{17}$

2. $\dfrac{7}{10} - \dfrac{5}{10}$

3. $1 - \dfrac{3}{4}$

4. $8\dfrac{5}{8} - 4\dfrac{2}{8}$

5. $\dfrac{36}{45} + \dfrac{4}{45}$

6. $\dfrac{8}{24} + \dfrac{9}{24} + \dfrac{7}{24}$

Evaluate $\dfrac{15}{22} - x$ for each value of x. Write your answers in simplest form.

7. $x = \dfrac{5}{22}$

8. $x = \dfrac{7}{22}$

9. $x = \dfrac{13}{22}$

Estimate each sum or difference by rounding to 0, $\dfrac{1}{2}$, or 1.

10. $\dfrac{3}{4} + \dfrac{3}{8}$

11. $\dfrac{7}{11} - \dfrac{2}{5}$

12. $\dfrac{4}{15} + \dfrac{7}{20}$

Answers: 1. $\dfrac{15}{17}$ 2. $\dfrac{1}{5}$ 3. $\dfrac{1}{4}$ 4. $4\dfrac{3}{8}$ 5. $\dfrac{8}{9}$ 6. 1 7. $\dfrac{5}{11}$ 8. $\dfrac{4}{11}$ 9. $\dfrac{1}{11}$ 10. 1 11. 0 12. $\dfrac{1}{2}$

Holt Mathematics

Name _____ Date _____ Class _____

CHAPTER 4 Family Fun
Three in a Row

Materials
Deck of cards
2 different colored pencils

Directions

- Play with a partner and take turns.

- At the beginning of each turn, each player draws 2 cards. The card with the lesser value is the numerator and the card with the greater value is the denominator. The face cards are wild and can be given any value.

- If your fraction is the answer to one of the problems in the box, then color that square. If both players have the same fraction, the player who finds the match first will be awarded that square.
  Hint: Do not forget about simplest form.

- The first player who colors three squares in a row is the winner.

- Create your own game if you wish in the space provided.

$\frac{5}{9} + \frac{1}{9}$	$\frac{3}{10} + \frac{2}{10}$	$\frac{6}{8} - \frac{4}{8}$
$\frac{14}{14} - \frac{8}{14}$	1	$\frac{1}{5} + \frac{3}{5}$
$\frac{8}{9} - \frac{1}{9}$	$\frac{2}{9} + \frac{3}{9}$	$\frac{12}{16} - \frac{10}{16}$

Answers: $\frac{6}{9}$ or $\frac{2}{3}$; $\frac{5}{10}$ or $\frac{1}{2}$; $\frac{2}{8}$ or $\frac{1}{4}$; $\frac{6}{14}$ or $\frac{3}{7}$; 1; $\frac{4}{5}$; $\frac{7}{9}$; $\frac{5}{9}$; $\frac{2}{16}$ or $\frac{1}{8}$

Family Letter

Chapter 5, Section A

What We Are Learning

Adding and Subtracting Fractions

Vocabulary
These are the math words we are learning:

least common denominator (LCD) the least common multiple of the denominators

least common multiple (LCM) the smallest number that is a multiple of two or more numbers

Dear Family,

In this section, your child will learn how to add and subtract fractions with unlike denominators. As a prelude to this skill, your child will first learn concepts that help him or her master this skill.

Your child will be learning the concept of **least common multiple** or **LCM**. The LCM is the smallest multiple of all the multiples that two or more numbers have in common. For example:

Find the least common multiple (LCM).
8, 12, and 16

8: 8, 16, 24, 32, 40, **48**, 56, 64, 72, 80, ...　List the multiples
12: 12, 24, 36, **48**, 60, 72, 84, 96, ...　　　of 8, 12, and 16.
16: 16, 32, **48**, 64, 80, 96, ...　　　　　　Find the smallest number that is in all the lists.

Notice that 48 is the least multiple that all three numbers have in common. So, the LCM of 8, 12, and 16 is 48.

Your child will learn to model fraction addition and subtraction. When a shape is divided into parts, each part represents a fraction of the whole. Your child can make drawings to represent each fraction in an addition or subtraction sentence and count the number of parts in each drawing to find the answer, as shown below.

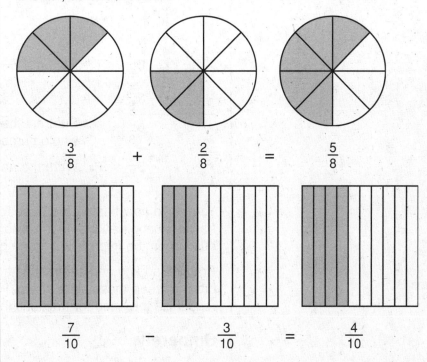

$$\frac{3}{8} + \frac{2}{8} = \frac{5}{8}$$

$$\frac{7}{10} - \frac{3}{10} = \frac{4}{10}$$

Holt Mathematics

Family Letter
Section A, continued

Being able to recognize the least common multiple of two or more numbers will help your child in finding the **least common denominator,** or **LCD,** of unlike fractions. Once the LCD is identified, your child can use it to find equivalent fractions that will help him or her find the sum or difference.

Add. Write your answer in simplest form.

$\frac{1}{4} + \frac{1}{6}$

Find the LCM of 4 and 6.
4: 4, 8, **12**, 16, 20
6: 6, **12**, 18, 24   LCM: 12

Since the LCM of 4 and 6 is 12, use this as the least common denominator of the two fractions to make equivalent fractions.

$\frac{1}{4} + \frac{1}{6}$ The LCD of the denominators is 12.

$\frac{3}{12} + \frac{2}{12}$ Write equivalent fractions.

$\frac{5}{12}$ Add.

As your child becomes more proficient in adding and subtracting fractions, the degree of difficulty of the problems will increase, such as with adding and subtracting mixed numbers.

Subtract. Write your answer in simplest terms.

$7\frac{1}{8} - 3\frac{5}{8}$

$7\frac{1}{8} \rightarrow 6\frac{9}{8}$ Rename $7\frac{1}{8}$ as $6 + 1\frac{1}{8} = 6 + \frac{8}{8} + \frac{1}{8}$.

$-3\frac{5}{8} \rightarrow -3\frac{5}{8}$

$3\frac{4}{8}$ Subtract the fractions and then the whole numbers.

$3\frac{4}{8} = 3\frac{1}{2}$ Write the answer in simplest form.

Your child now has a solid base for understanding fractions and fractional applications. In the following chapters, your child will build upon the information learned in this section.

Practice adding and subtracting fractions with your child in order to keep these skills sharp. Have your child begin to recognize the need for fractions in everyday situations.

Sincerely,

Name _____ Date _____ Class _____

CHAPTER 5
Family Letter
Adding and Subtracting Fractions

Model each problem and solve.

1. $\frac{6}{9} + \frac{2}{9}$
2. $\frac{5}{6} - \frac{1}{6}$
3. $\frac{3}{8} + \frac{5}{8}$
4. $\frac{6}{7} - \frac{5}{7}$

Find the least common multiple (LCM).

5. 4, 16

6. 3, 8

_____   _____

7. 2, 3, and 5

8. 5, 10, and 12

_____   _____

Add or subtract. Write each answer in simplest form.

9. $\frac{6}{7} + \frac{1}{2}$

10. $\frac{5}{9} - \frac{1}{3}$

_____   _____

11. $\frac{3}{8} + \frac{5}{12}$

12. $6\frac{3}{5} + 5\frac{1}{4}$

_____   _____

13. $5\frac{1}{9} + 8\frac{1}{3}$

14. $10\frac{6}{8} - 2\frac{1}{4}$

_____   _____

Find each sum or difference. Write the answer in simplest form.

15. $7 - 5\frac{3}{4}$

16. $6\frac{3}{5} + 4\frac{2}{3}$

17. $15\frac{1}{8} - 7\frac{5}{6}$

_____   _____   _____

Solve each equation. Write the solution in simplest form.

18. $y - 6\frac{1}{6} = 7\frac{1}{2}$

19. $x + 2\frac{4}{7} = 3\frac{1}{14}$

20. $3\frac{3}{8} = a - 8\frac{10}{16}$

_____   _____   _____

Answers: 1. + = = $\frac{8}{9}$ 2. − = $\frac{4}{6}$ or $\frac{2}{3}$ 3.  = $\frac{8}{8}$ or 1 4. 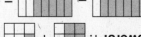 = $\frac{1}{7}$ 5. 16 6. 24 7. 30 8. 60 9. $1\frac{5}{14}$ 10. $\frac{2}{9}$ 11. $\frac{19}{24}$ 12. $11\frac{17}{20}$ 13. $13\frac{4}{9}$ 14. $8\frac{1}{2}$ 15. $1\frac{1}{4}$ 16. $11\frac{4}{15}$ 17. $7\frac{7}{24}$ 18. $y = 13\frac{2}{3}$ 19. $x = \frac{1}{2}$ 20. $a = 12$

39   Holt Mathematics

Name _____ Date _____ Class _____

CHAPTER 5 Family Fun
Magic Squares

Directions

- Find the solution to the magic squares. The sum of each row, column, and diagonal must be equal. In some instances you must finish the squares when the magic sum is given. Other times you need to determine the magic sum by finishing the square.
- Create your own magic square and see if your friends can find the solution.

1. The magic sum is $4\frac{4}{11}$.

		$1\frac{8}{11}$
$1\frac{9}{11}$		
	$1\frac{7}{11}$	$1\frac{6}{11}$

2. What is the magic sum? _____

	$2\frac{1}{6}$	
$2\frac{1}{3}$	$2\frac{7}{18}$	$2\frac{1}{9}$

3. The magic sum is $3\frac{15}{16}$.

		$1\frac{1}{4}$
$1\frac{1}{16}$		
	$1\frac{7}{16}$	$1\frac{1}{8}$

4. What is the magic sum? _____

		$2\frac{3}{5}$
	$1\frac{1}{2}$	
$\frac{2}{5}$		$1\frac{1}{5}$

Answers: 1. $\frac{4}{11}$, $1\frac{3}{11}$, $1\frac{5}{11}$, $1\frac{1}{11}$, $1\frac{2}{11}$ 2. $6\frac{6}{6}$, $2\frac{4}{9}$, $2\frac{2}{9}$, $2\frac{5}{18}$, $2\frac{1}{2}$ 3. $\frac{1}{2}$, $\frac{3}{16}$, $\frac{5}{16}$, $\frac{13}{16}$, $\frac{8}{16}$ 4. $\frac{1}{5}$, $\frac{3}{5}$, $4\frac{2}{5}$, $1\frac{4}{5}$ 1. $\frac{1}{10}$, $2\frac{3}{10}$, $\frac{7}{10}$, $2\frac{9}{10}$

Copyright © by Holt, Rinehart and Winston.
All rights reserved.

Holt Mathematics

CHAPTER 5 Family Letter
Section B

What We Are Learning

Multiplying and Dividing Fractions

Vocabulary
These are the math words we are learning:

reciprocal two numbers whose product equals 1

unlike fractions fractions with different denominators

Dear Family,

In this section your child will continue to study fractions. Your child will learn to multiply and divide fractions, as well as solve equations using fractions.

Your child has previously learned that multiplication is simply repeated addition. Using this knowledge, your child will learn to multiply fractions by whole numbers and write the answer in simplest form.

Your child will need to remember that whole numbers can be rewritten as improper fractions with a 1 in the denominator. For example, $7 = \frac{7}{1}$ or $25 = \frac{25}{1}$.

To multiply fractions, your child will learn to multiply the numerators *and* the denominators. He or she will then write the product in simplest form.

Multiply. Write the answer in simplest form.

$\frac{3}{4} \cdot \frac{2}{5}$

$\frac{3}{4} \cdot \frac{2}{5} = \frac{3 \cdot 2}{4 \cdot 5}$ Multiply the numerators. Multiply the denominators.

$= \frac{6}{20}$ The GCF of 6 and 20 is 2.

$= \frac{3}{10}$ Write the answer in simplest form.

To multiply fractions and mixed numbers, your child will learn to first change the mixed number to an improper fraction. Once the mixed number is in this new form, your child will multiply the numerators and the denominators.

Multiply. Write the answer in simplest form.

$2\frac{1}{4} \cdot 1\frac{2}{3}$

$\frac{9}{4} \cdot \frac{5}{3}$ Write the mixed numbers as improper fractions.

$\frac{9 \cdot 5}{4 \cdot 3} = \frac{45}{12}$ Multiply numerators. Multiply denominators.

$= 3\frac{9}{12}$ You can write the improper fraction as a mixed number.

$= 3\frac{3}{4}$ Simplify.

Holt Mathematics

Family Letter

Chapter 5 Section B, continued

Your child will use concepts from multiplying fractions to successfully divide fractions. Your child will use the **reciprocal** of a fraction when dividing fractions. To find the reciprocal of a number, simply flip or switch the numerator and the denominator of the fraction or whole number. The product of a number and its reciprocal is always 1.

Find the reciprocal.

$\frac{6}{7}$

$\frac{6}{7} \cdot ? = 1$ Think: $\frac{6}{7}$ of what number is 1?

$\frac{6}{7} \cdot \frac{7}{6} = 1$ $\frac{6}{7}$ of $\frac{7}{6}$ is 1.

The reciprocal of $\frac{6}{7}$ is $\frac{7}{6}$.

Dividing by a number is the same as multiplying by its reciprocal.

Divide. Write the answer in simplest form.

$\frac{2}{3} \div \frac{1}{4}$

$\frac{2}{3} \div \frac{1}{4} = \frac{2}{3} \cdot \frac{4}{1}$ Rewrite as multiplication using the reciprocal of $\frac{1}{4}$, 4.

$= \frac{2 \cdot 4}{3 \cdot 1} = \frac{8}{3}$ Multiply by the reciprocal.

$= 2\frac{2}{3}$ You can write the answer as a mixed number.

Your child will apply the skills from this section when asked to solve equations, particularly equations that contain fractions. Multiplying and dividing fractions are important tools that your child will use throughout this course. Review these skills daily with your child.

Sincerely,

CHAPTER 5 Family Letter
Adding and Subtracting Fractions

Multiply. Write your answers in simplest form.

1. $6 \cdot \frac{1}{15} =$ _____
2. $9 \cdot \frac{1}{20} =$ _____
3. $3 \cdot \frac{3}{25} =$ _____

Evaluate 9x for each value of x. Write your answers in simplest form.

4. $x = \frac{1}{3}$ _____
5. $x = \frac{5}{9}$ _____
6. $x = \frac{7}{9}$ _____

Multiply. Write the answer in simplest form.

7. $\frac{1}{2} \cdot \frac{3}{5} =$ _____
8. $\frac{1}{4} \cdot \frac{6}{7} =$ _____
9. $\frac{5}{9} \cdot \frac{3}{10} =$ _____
10. $\frac{2}{3} \cdot \frac{3}{8} =$ _____

Evaluate the expression $x \cdot \frac{1}{3}$ for each value of x.

11. $x = \frac{3}{8}$ _____
12. $x = \frac{9}{10}$ _____
13. $x = \frac{6}{11}$ _____

Multiply. Write the answer in simplest form.

14. $3\frac{1}{4} \cdot \frac{1}{8} =$ _____
15. $5 \cdot 2\frac{1}{5} =$ _____
16. $2\frac{1}{3} \cdot 3\frac{1}{2} =$ _____

Find the reciprocal.

17. $\frac{6}{9}$ _____
18. $\frac{7}{11}$ _____
19. $\frac{6}{7}$ _____
20. $\frac{1}{2}$ _____

Divide. Write the answer in simplest form.

21. $\frac{7}{9} \div 3$
22. $\frac{4}{7} \div \frac{5}{7}$
23. $3\frac{1}{2} \div 1\frac{7}{8}$
24. $\frac{8}{15} \div 2\frac{3}{5}$

Solve each equation. Write the answer in simplest form.

25. $5x = \frac{2}{3}$ _____
26. $\frac{2}{5}x = 18$ _____
27. $\frac{8x}{9} = 16$ _____

Answers: 1. $\frac{2}{5}$ 2. $\frac{9}{20}$ 3. $\frac{9}{25}$ 4. 3 5. 5 6. 7 7. $\frac{3}{10}$ 8. $\frac{3}{14}$ 9. $\frac{1}{6}$ 10. $\frac{1}{4}$ 11. $\frac{1}{8}$ 12. $\frac{3}{10}$ 13. $\frac{2}{11}$ 14. $\frac{13}{32}$ 15. 11 16. $\frac{49}{6}$ 17. $\frac{3}{2}$ 18. $\frac{11}{7}$ 19. $\frac{7}{6}$ 20. 2 21. $\frac{7}{27}$ 22. $\frac{4}{5}$ 23. $1\frac{13}{15}$ 24. $\frac{8}{39}$ 25. $x = \frac{2}{15}$ 26. $x = 45$ 27. $x = 18$

Family Fun
Four in a Row Fractions

CHAPTER 5

Directions

- Cut out and shuffle the problem cards below.
- Each player should create a 4-by-4 grid as shown.
- Have each player write the fractions from the shaded box below in any order in each one of the squares in the grid.
- Taking turns, each player will draw a card from the problem cards, and find the product or quotient in simplest terms.
- If the player has the simplified fraction on his or her grid, he or she will color in the corresponding square.
- The first player to color in four squares in a row, column, or diagonal line wins the game.

$\frac{1}{4}$ $\frac{5}{12}$ $\frac{1}{2}$ $\frac{10}{27}$ $1\frac{2}{3}$ $\frac{4}{21}$ $1\frac{1}{2}$ $\frac{7}{72}$ $\frac{1}{10}$ $\frac{4}{5}$ $\frac{13}{25}$ $\frac{1}{14}$ $\frac{27}{40}$ $\frac{1}{22}$ $\frac{27}{32}$ $\frac{2}{11}$

$\frac{5}{12} \times \frac{3}{5}$	$\frac{1}{3} \div \frac{4}{5}$	$\frac{3}{5} \times \frac{5}{6}$	$\frac{5}{9} \times \frac{2}{3}$
$8 \div 4\frac{4}{5}$	$\frac{6}{7} \times \frac{2}{9}$	$\frac{3}{8} \div \frac{1}{4}$	$\frac{1}{9} \times \frac{7}{8}$
$\frac{9}{10} \times \frac{1}{9}$	$\frac{8}{15} \div \frac{2}{3}$	$\frac{13}{15} \times \frac{3}{5}$	$\frac{2}{7} \times \frac{1}{4}$
$\frac{3}{8} \div \frac{5}{9}$	$\frac{1}{10} \times \frac{5}{11}$	$\frac{3}{4} \div \frac{8}{9}$	$\frac{3}{7} \times \frac{14}{33}$

Holt Mathematics

Family Letter
Chapter 6, Section A

What We Are Learning

Organizing Data

Vocabulary
These are the math words we are learning:

mean the sum of all the items, divided by the number of items in the set (sometimes called an average)

median the middle value when the data are in numerical order or the mean of the two middle numbers if there is an even number of items

mode the value or values that occur most often in a set of data

outlier a value in a set that is very different from the other values and may affect the mean

range the difference between the least and the greatest values in the set

Dear Family,

Your child is beginning to study statistics and the different ways that data can be represented and organized. In this section, your child will use tables to record and organize data.

By organizing data in a table, your child will be able to easily see patterns and relationships, which will help him or her formulate sensible conclusions.

Use the temperature data to make a table. Then use your table to find a pattern in the data and draw a conclusion.

At 6 A.M. the temperature was 15°F; at 10 A.M. it was 22°F; at 2 P.M. it was 35°F; at 6 P.M. it was 21°F; at 10 P.M. it was 10°F.

Time	Temperature
6 A.M.	15°F
10 A.M.	22°F
2 P.M.	35°F
6 P.M.	21°F
10 P.M.	10°F

By organizing the temperature data in the table, it is easy to see that during the time span noted, the temperature rose every four hours until 2 P.M. After 2 P.M. the temperature started to fall every four hours.

One conclusion that can be made from the data in the table is that the low temperature for the day was at least 10° F. See if your child can make another conclusion.

Making a table to organize data is an important problem-solving strategy. A table can help your child decipher confusing data in a clear and concise way. Tables are also helpful when trying to determine if the data has a pattern or when predicting other results.

Family Letter
Chapter 6, Section A, continued

Once your child has organized data, he or she can use the mean, median, mode, and range to describe it.

Mean	The sum of the values divided by the number of values, otherwise known as the average of the numbers.
Median	Once the values are organized numerically from least to greatest, the median is the middle value.
Mode	The value or values that occur most often in a set of data. If no value occurs more than once, then there is no mode.
Range	The difference between the greatest value and the least value.

Find the range, mean, median, and mode of the data set.
45, 44, 48, 42, 12, 44, 45

Range Subtract the least value from the greatest value.
48 − 12 = 36

The range is 36.

Mean Add the values. Divide the sum by the number of items.
45 + 44 + 48 + 42 + 12 + 44 + 45 = 280
280 ÷ 7 = 40

The mean is 40.

Median Arrange the values in order. Choose the number in the middle of the data set.
12, 42, 44, 44, 45, 45, 48

The median is 44.

Mode The values 44 and 45 each occur 2 times.

The modes are 44 and 45.

Your child will also learn to identify an **outlier**. An outlier is an extreme piece of data. Ask your child how an outlier can affect the mean, median, mode, and range of a data set.

Displaying and analyzing data has an important role in mathematics. Encourage your child to be aware of statistics in everyday situations.

Sincerely,

Name _____ Date _____ Class _____

CHAPTER 6 Family Letter
Organizing Data

1. Ms. Fike has been teaching for five years. Her first year, she had 22 students. Her second year, she had 25 students. Her third year, she had 28 students. Her fourth year, she had 31 students. Her fifth year, she had 34 students. Use this data to make a table.

2. Use your table from Exercise 1 to find a pattern in the data and draw a conclusion.

_____

_____

Find the range, mean, median, and mode of each data set.

Price of computer games	$34	$28	$57	$30	$38	$69	$44	$30	$30

3. Mean

_____

4. Median

_____

5. Mode

_____

6. Range

_____

ATTENDANCE AT SCOTT MIDDLE SCHOOL						
Year	1997	1998	1999	2000	2001	2002
Attendance	180	195	210	192	185	208

7. Mean

_____

8. Median

_____

9. Mode

_____

10. Range

_____

11. Kate sells kitchen tools at a local kitchen store. Her sales for the past 5 weeks were $856, $1,034, $798, $2,950, and $832. What are the mean, median, and mode of the data? Which one best describes the data set?

_____

Answers: 1. 1: 22; 2: 25; 3: 28; 4: 31; 5: 34. 2. The class is increasing by 3 each year. The school population is increasing each year. 3. $40 4. $34 5. $30 6. $41 7. 195 8. 193.5 9. no mode 10. 30 11. The mean is $1,294; There is no mode; The median is $856; The median best describes the data because the value, $2,950, is an outlier.

Copyright © by Holt, Rinehart and Winston.
All rights reserved.

Holt Mathematics

Name _____ Date _____ Class _____

CHAPTER 6 Family Fun
Just a Mean Game

Materials

- Paper in different colors or patterns
- Centimeter ruler
- Bag
- Scissors

Directions

- Draw rectangular strips on paper. All the strips should be 1 cm wide. Vary the length—1 cm, 2 cm, 3 cm, and so on, up to 10 cm. Cut several strips of each length. Cut all strips of the same length from the same color. For example, cut all the 1 cm strips from yellow paper. Make at least 30 strips in all.
- Put the strips in a bag and mix them up. Take out a handful of strips.
- Work with a partner to arrange the strips in order from shortest to longest.
- Use the table to record the length of each strip.
- Find the mean, median, mode, and range of the strips you selected.

Length of Strips (cm)

	1	2	3	4	5	6	7	8	9	10
Number of Strips										

Mean _____

Median _____

Mode _____

Range _____

CHAPTER 6 Family Letter
Section B

What We Are Learning

Displaying and Interpreting Data

Vocabulary
These are the math words we are learning:

bar graph displays data with vertical or horizontal bars

coordinate grid a plane formed by the intersection of the *x*-axis and the *y*-axis

double-bar graph a bar graph that shows two related sets of data

double-line graph line graphs that display two sets of data

frequency table a table that tells the number of times an event, category, or group occurs

histogram a bar graph that shows the number of data items that occurs within each interval

line graph displays a set of data using line segments

line plot a method of displaying data that uses a number line to show the frequency of values

ordered pair two numbers that name a location on a coordinate grid

stem-and-leaf plot shows data arranged by place value

Dear Family,

In this section, your child will be interpreting and constructing many different types of graphs used to display data. Some of the graphs your child will be creating will be bar graphs, double bar graphs, line graphs, double line graphs, histograms, and stem-and-leaf plots.

Bar graphs and double bar graphs are helpful to display data that can be put into categories.

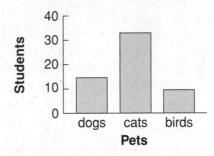

Line graphs and **double line graphs** are useful when displaying data over a period of time.

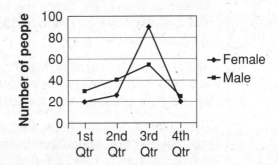

Plotting points on a line graph will prepare your child for graphing ordered pairs on a coordinate grid. The first number in an ordered pair tells you how far to move horizontally from the point (0, 0). The second number in an ordered pair tells you how far to move vertically.

Plotting points is an important skill that will help to prepare your child for future mathematical concepts.

Copyright © by Holt, Rinehart and Winston.
All rights reserved.

Holt Mathematics

Family Letter
Section B, continued

There are many different ways to represent data. Sometimes a graph can display data in a misleading way. Misleading graphs rely on someone looking at the shape of the graph, not what the graph is actually representing. Your child will learn to focus on the scale of the graph to be sure that the data is accurately represented.

Explain why this bar graph is misleading.

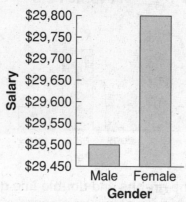

This graph is misleading because it looks as if females earn a lot more than males, while there is actually only a $300 difference in salaries. Since the scale does not start at zero and the intervals increase by $50, the difference in salaries appears to be very large.

Search through magazines and newspapers with your child to find graphs that display misleading data and graphs that accurately represent the data presented.

The study of data is a topic that will be covered throughout your child's math education. Be sure to be involved as your child continues to build a foundation for a great math future.

Sincerely,

Name _____ Date _____ Class _____

Family Letter
CHAPTER 6
Displaying and Interpreting Data

Use the double bar graph to answer each question.

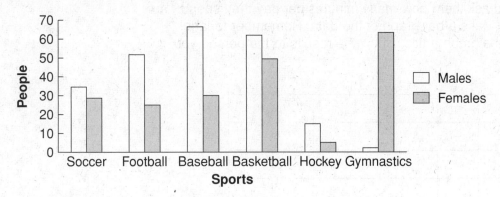

1. Which sport is liked best by males?
   _____

2. Which sport do both males and females prefer to watch?
   _____

Explain why this line graph is misleading.

3. _____
   _____
   _____
   _____
   _____

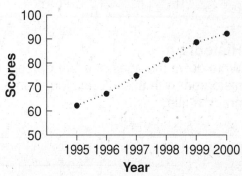

Math Scores for Brookpark

Use the data in the table to complete the stem-and-leaf plot.

4.
Stem	Leaves
6	
7	
8	
9	

Key: 6|6 means 66

85	99	65	78	85	82	83
66	74	92	79	92	84	99
80	75	89	92	86	70	65

Answers: 1. Baseball 2. Basketball 3. The scale does not start at zero. It looks as if the scores dramatically increased over 6 years. 4. 6 | 5 5 6; 7 | 0 4 5 8 9; 8 | 0 2 3 4 5 5 6 9; 9 | 2 2 2 9 9

51 Holt Mathematics

Name _____ Date _____ Class _____

CHAPTER 6 Family Fun
Graph It

Bar Graph

Make a list of activities that most people do on a daily basis, such as climbing stairs, walking to the car, vacuuming, etc. Interview a person in your household and ask them how many minutes per day they spend doing each activity. Make a bar graph of the data. Remember to label the axes and give the graph a title. Show the results to the person you interviewed.

Histogram

Write down the ages of every person in your household or family. Create a histogram with this information. Remember to label the axes and give the graph a title.

Holt Mathematics

Family Letter
Chapter 7, Section A

What We Are Learning

Ratios and Proportions

Vocabulary
These are the math words we are learning:

equivalent ratios ratios that name the same comparison

corresponding angles congruent angles in similar figures

corresponding sides sides of similar figures whose lengths are proportional to each other

indirect measurement the use of similar figures and proportions to find the measure of an object that cannot be directly measured

proportion an equation that states two ratios are equivalent

rate a comparison of two quantities that have different units of measure

ratio the comparison of two quantities by division

scale the ratio between two sets of measurements

scale drawing a drawing of a real object that is proportionally smaller or larger than the real object

Dear Family,

Your child will be learning how to write and solve problems involving ratios, rates, and proportions. A **ratio** compares two items or quantities using division and can be written in one of three ways: comparing a part to a part, a part to a whole, or a whole to a part.

Use the table to write each ratio.

Boys	Girls
11	12

boys to girls

$\frac{11}{12}$ or 11 to 12 or 11:12   part to part

boys to total number of students

$\frac{11}{23}$ or 11 to 23 or 11:23   part to whole

total number of students to girls

$\frac{23}{12}$ or 23 to 12 or 23:12   whole to part

A **rate** compares two quantities that have different measures. By learning to find the **unit rate** between two items, your child will be able to determine which rate is the better deal.

Grocery Store A is selling 5 pounds of potatoes for $3.50. Grocery Store B is selling 3 pounds of potatoes for $1.89. Which is the better deal?

Grocery Store A		Grocery Store B	
$\frac{\$3.50}{5 \text{ lb}}$	Write the rate.	$\frac{\$1.89}{3 \text{ lb}}$	Write the rate.
$\frac{\$3.50 \div 5}{5 \text{ lb} \div 5}$	Divide both terms by 5.	$\frac{\$1.89 \div 3}{3 \text{ lb} \div 3}$	Divide both terms by 3.
$\frac{\$0.70}{1 \text{ lb}}$	$0.70 per pound	$\frac{\$0.63}{1 \text{ lb}}$	$0.63 per pound

Grocery Store B has the better deal on the potatoes. The unit rate of $0.63 per pound is cheaper than that of $0.70 per pound.

Family Letter
Chapter 7 — Section A, continued

similar figures that have exactly the same shape but not necessarily the same size

unit rate a rate in which the second quantity in the comparison is one unit

Your child will apply the concepts of ratios to solve proportions. A **proportion** is an equation that states two ratios are equivalent. Proportions are very helpful in everyday life situations. Your child will use proportions to make conversions between units, identify similar figures, find indirect measurements, and determine distances with maps and scales.

This is how your child will use cross products to find a missing value in a proportion.

Find the missing value in the proportion.

$$\frac{3}{8} = \frac{9}{x}$$

$\frac{3}{8} \times \frac{9}{x}$ Find the cross products

$3 \cdot x = 8 \cdot 9$ The cross products are equal.
$3x = 72$

$\frac{3x}{3} = \frac{72}{3}$ Divide both sides by 3 to undo the multiplication.

$x = 24$

Your child can also use proportions to make conversions within the customary measurement system. An example of this is shown below.

Dave has 63 feet of material. How many yards can be cut from the material?

One yard equals 3 feet, so use the unit conversion factor

$\frac{1 \text{ yard}}{3 \text{ feet}}$ or $\frac{3 \text{ feet}}{1 \text{ yard}}$.

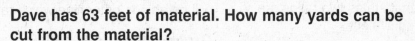
$63 \text{ ft} \cdot \frac{1 \text{ yd}}{3 \text{ ft}} = \frac{63}{3} = 21 \text{ yd}$ Multiply and cross out "ft".

Dave can cut 21 yards from the material.

By knowing how to find missing values using proportions, your child will be able to solve a myriad of real life math problems. Have your child tell you how to find a missing value when solving problems that use proportions.

Sincerely,

Name _____ Date _____ Class _____

CHAPTER 7 Family Letter
Ratios and Proportions

Use the table to write each ratio.

	Mr. Worth's Class	Mrs. Adam's Class
6th graders	28	24
7th graders	18	26

1. 6th graders to 7th graders

2. Mrs. Adam's 7th graders to Mr. Worth's 6th graders

3. Company X sells a 64-oz carton of orange juice that costs $2.24. Company Y sells a 48-oz carton of orange juice for $1.92. Which is the better deal?

Find the missing value in each proportion.

4. $\frac{3}{4} = \frac{x}{16}$

5. $\frac{5}{14} = \frac{25}{m}$

6. $\frac{6}{x} = \frac{18}{30}$

7. St. Paul's Cathedral has a tower that is 366 ft tall. How many yards is this?

8. The tallest hand-built sand sculpture took 96 hours to build. How many days is this?

9. A picture is 10 inches long by 8 inches wide. A similar picture is 2.5 inches long. How wide is it?

10. A tree casts a shadow 30 ft long. At the same time, a 5-ft woman casts a 3 ft shadow. How tall is the tree?

11. On a city map, the school is 1.75 cm from the park. If the scale is 1 cm = 6 km, what is the actual distance from the school to the park?

Answers: 1. $\frac{13}{11}$ 2. $\frac{13}{14}$ 3. 64-oz carton 4. $x = 12$ 5. $m = 70$ 6. $x = 10$ 7. 122 yards 8. 4 days 9. 2 in. 10. 50 ft tall 11. 10.5 km

Name _____ Date _____ Class _____

CHAPTER 7 Family Fun
Smart Shoppers

A smart shopper can find the best value on a certain product by comparing unit prices. Practice being a smart shopper by circling the better deal for each item listed below.

Item	Bob's Bargains	Dottie's Deals
Pet Food	15 pounds for $16.50	10 pounds for $12.99
Shampoo	24 ounces for $0.99	16 ounces for $0.75
Laundry Detergent	64 ounces for $3.99	128 ounces for $6.99
Printer Ink	4 cartons for $90.00	2 cartons for $42.00
Apples	5 pound bag for $3.99	3 pound bag for $1.75
Milk	1 gallon for $2.59	3 gallons for $8.00
Pencils	24 for $2.09	12 for $0.99
Eggs	1 dozen for $1.09	18 eggs for $2.00

Answers: Bob, Bob, Dottie, Dottie, Bob, Dottie, Bob, Bob

CHAPTER 7 Family Letter
Section B

What We Are Learning

Percents

Vocabulary
These are the math words we are learning:

discount the amount that is subtracted from the original price of an item

percent a ratio comparing a number to 100

sales tax a percent of the cost of an item, which is charged by states to raise money

tip the amount of money added to a bill

Dear Family,

Your child will be learning about the relationship between percents, decimals, and fractions. In order to be able to recognize this relationship, your child will learn how to write a percent as a decimal and as a fraction.

Your child will also be converting from decimals and fractions to percents. Your child will learn that a **percent** is a ratio comparing a number to 100 and is written using the % symbol.

Percents can be written as fractions or decimals. Likewise, both fractions and decimals can be written as percents.

$$25\% \quad = \quad \frac{25}{100} \quad = \quad 0.25$$

Percent   Fraction   Decimal

Your child can model percents in much the same way as fractions and decimals are modeled, by shading in the proper portion of a 10 by 10 grid. For example:

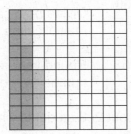

25%

Write 60% as a fraction in simplest form.

$60\% = \frac{60}{100}$ Write the percent as a fraction with a denominator of 100.

$\frac{60}{100} = \frac{3}{5}$ Write the fraction in simplest form.

Write $\frac{7}{8}$ as a percent.

$\frac{7}{8} = 7 \div 8 = 0.875$ Use division to write the fraction as a decimal.

$0.875 = 87.5\%$ Write the decimal as a percent.

Write 0.89 as a percent.

$0.89 = \frac{89}{100}$ Use what you know about place value to express the decimal as a fraction (eighty-nine hundredths).

$\frac{89}{100} = 89\%$ Write the numerator with a percent symbol.

Holt Mathematics

Family Letter

Section B, continued

Once your child has mastered converting between percents, decimals, and fractions, he or she will begin to solve problems using percents. Your child will learn to set up proportions and then find the missing value. Here is an example of how your child will use proportions to solve percent problems.

There are 650 students at Highland Elementary School. Thirty-five percent of all students attending the school ride the bus. Find the number of students that ride the bus.

$\dfrac{\%}{100} = \dfrac{is}{of}$ Set up the proportion.

$\dfrac{35}{100} = \dfrac{s}{650}$ Let s equal the number of students who ride the bus. Set up a proportion. You need to find the "is" part of the proportion.

$100s = 22{,}750$ The cross products are equal.

$\dfrac{100s}{100} = \dfrac{22{,}750}{100}$ Divide both sides by 100 to undo the multiplication.

$s = 227.5$ Round your answer to the nearest whole number so your answer will be reasonable.

There are about 228 students who ride the bus each day to school.

Percents are a valuable part of any math program. It allows students to apply what they learn to real life situations. Encourage your child to calculate the tax on a purchase or the tip for services rendered. Seeing math used in everyday life will have a profound effect on your child's math performance.

Sincerely,

Name _____ Date _____ Class _____

CHAPTER 7
Family Letter
Percents

Write each percent as a fraction in simplest form.

1. 60%   2. 12%   3. 51%   4. 7%

_____   _____   _____   _____

Write each percent as a decimal.

5. 8%   6. 43%   7. 99%   8. 12%

_____   _____   _____   _____

Write each decimal as a percent.

9. 0.462   10. 0.2   11. 0.83   12. 0.05

_____   _____   _____   _____

Write each fraction as a percent.

13. $\frac{5}{8}$ 14. $\frac{2}{5}$ 15. $\frac{39}{50}$ 16. $\frac{9}{20}$

_____   _____   _____   _____

Solve.

17. Joe owns 30 music CDs. If rap music makes up 40% of his collection, how many rap CDs does Joe own?

_____

18. Mrs. Gonzalez ordered a dining room table. She had to pay 20% of the total cost when she ordered the table. She will pay the remaining balance when the table is delivered. If she has already paid $60, how much more does she owe?

_____

19. Find 66% of 250.

_____

20. Jean went out to dinner. Her bill was $16.71. If she left a tip that was 15% of the bill, about how much was the amount of the tip?

_____

Answers: 1. $\frac{3}{5}$ 2. $\frac{3}{25}$ 3. $\frac{51}{100}$ 4. $\frac{7}{100}$ 5. 0.08 6. 0.43 7. 0.99 8. 0.12 9. 46.2% 10. 20% 11. 83% 12. 5% 13. 62.5% 14. 40% 15. 78% 16. 45% 17. 12 CDs 18. $240 19. 165 20. $2.50

59   Holt Mathematics

Name _____ Date _____ Class _____

CHAPTER 7 — Family Fun
Let's Go Shopping!

You have just won $500 in a math contest! Your parents want you to put at least 30% of the prize money in your college savings account. You can spend rest of the money however you wish.

Listed below are some items on sale at the local mall. Use the chart to help you calculate the sale price for each item. Next choose which items you will purchase. Don't forget the 5% sales tax on your entire purchase. Once you are finished shopping, don't forget to stop by the bank to make the deposit into your saving account!

Item	Original Price	Sale	Sale Price
Computer Game	$55.00	30% off	
Basketball	$15.00	15% off	
Jeans	$32.00	25% off	
Sweater	$35.00	35% off	
Book	$12.50	50% off	
Earrings	$22.00	10% off	
Television	$268.00	15% off	
Stereo	$99.00	20% off	
Printer	$215.00	25% off	

Items you purchased. _____
_____

Total cost of your purchases (including tax). _____

Amount of money you saved on your purchases. _____

Amount of money you put into your college savings account. _____

Answers: Computer Game $38.50, Basketball $12.75, Jeans $24, Sweater $22.75, Book $6.25, Earrings $19.80, Television $227.80, Stereo $79.20, Printer $161.25; The amount in the savings account is at least $150.

Family Letter
Chapter 8, Section A

What We Are Learning

Lines and Angles

Vocabulary
These are the math words we are learning:

acute angle an angle that measures less than 90°

adjacent angles angles that are side by side and have a common vertex and ray

angle a figure formed by two rays with a common endpoint called a vertex

complementary angles two angles whose measures add to 90°

congruent angles that have the same measure

line a set of points that extends without end in opposite directions

line segment a part of a line with two endpoints

obtuse angle an angle whose measure is greater than 90° but less than 180°

parallel lines lines that are in the same plane and never intersect

perpendicular lines lines that intersect to form right angles

plane a flat surface that extends without end in all directions

Dear Family,

In this section, your child will be learning the necessary terminology for understanding and applying the concepts of geometry. When given a figure, your child will be able to identify **points, segments, rays, lines,** and **planes.** Use the chart to review of these basic concepts:

Concept	Definition	Symbols
Point	An exact location in space.	Named by a capital letter, *P*.
Line	A set of points that extends in both directions without end.	Named by two points on the line, $\overleftrightarrow{AB}$.
Ray	A part of a line that has one endpoint and extends forever.	Named by the endpoint first and then another point on the line, $\overrightarrow{AB}$.
Line Segment	A part of a line with two endpoints.	Named by the endpoints, $\overline{AB}$.
Plane	A flat surface that extends without end in all directions.	Named by three points on the plane that are not on a line, plane *ABC*.

Ask your child to explain the difference between a line, a ray, and a segment and how to differentiate between the symbols that are used to describe each of these terms in a given figure.

Family Letter
Chapter 8 — Section A, continued

point an exact location in a plane

ray a part of a line that has one endpoint and extends forever

right angle an angle that measures exactly 90°

skew lines lines that lie in different planes that are neither intersecting nor parallel

straight angle an angle that measures exactly 180°

supplementary angles two angles whose measures add to 180°

vertex the common endpoint formed by two rays

vertical angles a pair of opposite congruent angles formed by intersecting lines

Now that your child has a basic understanding of geometric terms, he or she will begin to apply this knowledge to other geometric concepts. One concept includes angles and angle relationships. Your child will learn to recognize and classify angles into four different categories: **acute, obtuse, right,** and **straight**.

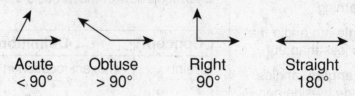

Acute   Obtuse   Right   Straight
< 90°   > 90°   90°   180°

Angles and angle relationships are an important part of understanding geometry. Two angles whose measures add to 90° are called **complementary angles.** Two angles whose measures add to 180° are called **supplementary angles.**

When two angles are opposite of each other, as when two lines intersect, **vertical angles** are created. These angles are said to be **congruent,** or have the same measure. By knowing these basic relationships among angles, your child will be able to find missing angle measures.

Identify a pair of vertical angles.

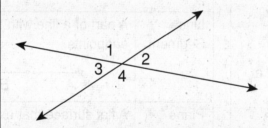

∠1 and ∠4 are opposite each other and are formed by intersecting lines. They are congruent vertical angles.

This is just an overview of what your child will be learning in this section. The information covered here will provide a strong foundation for the concepts and applications your child will be using throughout this chapter and future mathematic courses.

Sincerely,

Name _____ Date _____ Class _____

Family Letter
Lines and Angles

CHAPTER 8

Use the diagram to name each geometric figure.

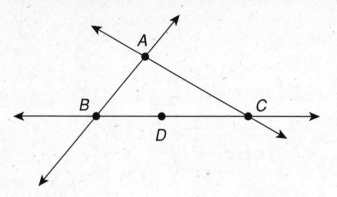

1. two points
_____

2. a plane
_____

3. three segments
_____

Use a protractor to measure each angle. Tell what type of angle it is.

4.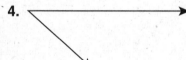
_____

5.
_____

Find each unknown angle measure.

6. The angles are complementary.

_____

7. The angles are vertical.

_____

Classify the pair of lines.

8.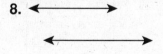
_____

9.
_____

Answers: 1. A, B, C, or D 2. Any three letters, such as ABC 3. $\overline{AB}$, $\overline{BD}$, $\overline{DC}$, $\overline{BC}$, or $\overline{AC}$ 4. 45°; acute 5. 90°; right 6. 38° 7. 110° 8. parallel 9. intersecting

63

Holt Mathematics

Family Fun
Crossword Geometry

Across
3. Angle that measures more than 90°
9. Lines that intersect to form 90° angles
12. Lines in the same plane that will never meet
14. Angles with the same angle measure
16. Two angles whose measures add to 90°
17. Angle that measures exactly 180°
19. A straight path that extends without end in opposite directions

Down
1. A line _____ is made up of two endpoints
2. Lines that are not parallel and will never intersect
4. Angle less than 90°
5. Formed by two rays
6. Angles that are always congruent
7. An exact location
8. The common point of an angle
10. Only has one endpoint
11. A flat surface that extends without end in all directions
13. Share a common vertex and ray
15. Two angles whose measures add to 180°
18. An angle that is exactly 90°

Answers: Across 3. obtuse angle 9. perpendicular 12. parallel 14. congruent 16. complementary 17. straight angle 19. line Down: 1. segment 2. skew 4. acute angle 5. angle 6. vertical angles 7. point 8. vertex 10. ray 11. plane 13. adjacent angles 15. supplementary 18. right angle

Chapter 8 Family Letter
Section B

What We Are Learning

Polygons

Vocabulary
These are the math words we are learning:

acute triangle a triangle with all angles less than 90°

equilateral triangle a triangle with three congruent sides

isosceles triangle a triangle with at least two congruent sides

obtuse triangle a triangle containing one obtuse angle

parallelogram a quadrilateral with two pairs of parallel sides

polygon a closed plane figure formed by three or more line segments that intersect only at their endpoints

quadrilateral a four-sided polygon

rectangle a parallelogram with four right angles

regular polygon a polygon with congruent sides and angles

rhombus a parallelogram with congruent sides

right triangle a triangle with a right angle

scalene triangle a triangle with no congruent sides

Dear Family,

In this section, your child will be learning the properties of various triangles and other polygons. She or he will also learn the relationship between these geometric figures.

Triangles are special **polygons** that are closed figures with three sides and three angles. The sum of the angle measures in any triangle is 180°. Like angles, triangles have special names and characteristics. Triangles can be classified by their angles or by the length of their sides.

Acute Triangle	△	All angles measure less than 90°.
Right Triangle	◣	There is one right angle.
Obtuse Triangle	◁	There is one angle measuring greater than 90°.
Isosceles Triangle	△	The lengths of two sides are congruent.
Equilateral Triangle	△	The lengths of all three sides are congruent.
Scalene Triangle	◿	The lengths of the sides are not congruent.

Your child will use the information about the different triangles to classify triangles and to help find the measure of an angle or the length of a side.

Family Letter
Section B, continued

square a rectangle with four congruent sides

trapezoid a quadrilateral with exactly one pair of parallel sides

Quadrilaterals are another type of special polygon. Quadrilaterals are four-sided closed figures. Some quadrilaterals are unique and have special names and characteristics. Your child will learn to identify and classify given quadrilaterals based on the following properties.

QUADRILATERALS
Any polygon with 4 sides and 4 angles

PARALLELOGRAM
A quadrilateral with two pairs of parallel sides

TRAPEZOID
A quadrilateral with exactly one pair of parallel sides

RECTANGLE
A parallelogram with 4 right angles

RHOMBUS
A parallelogram with 4 congruent sides

SQUARE
A rectangle that has four congruent sides or a rhombus with four right angles

Triangles and quadrilaterals are just two types of polygons your child will study in this section. Your child will also study other polygons, and learn to identify them as being either regular or irregular.

Knowing the basic figures of geometry and their properties are necessary in the study of geometry. Review the different figures and their special properties with your child to help with mastery of these concepts.

Sincerely,

Name _____ Date _____ Class _____

CHAPTER 8 Family Letter
Polygons

Use the diagram to find the measure of each indicated angle.

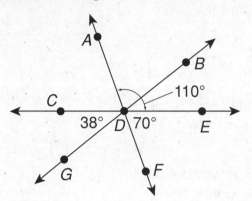

1. ∠CDA

2. ∠BDE

3. ∠ADB

_____   _____   _____

Classify the triangle using the given information.

4. The sum of the lengths of the three sides is 36 inches.

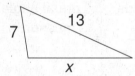

_____

Give the most descriptive name for each figure.

5.

6.

7.

_____   _____   _____

Complete each statement.

8. A parallelogram with four right angles is a _____.

9. A quadrilateral with exactly one pair of parallel sides is a _____.

Name each polygon and tell whether it appears to be *regular* or *not regular*.

10.

11. (hexagon)

12. (pentagon)

_____   _____   _____

Answers: 1. 70° 2. 38° 3. 72° 4. scalene 5. square or rhombus 6. trapezoid 7. parallelogram 8. rectangle 9. trapezoid 10. not regular quadrilateral 11. regular hexagon 12. not regular pentagon

Name _____ Date _____ Class _____

CHAPTER 8 Family Fun
Draw It!

Directions
The goal is for your team to be the first team to correctly draw the given polygon.

- Pick someone to be the announcer and the judge.
- Form teams of two people.
- Cut out the cards. The announcer picks a card from the deck and reads the description.
- Each team has to name and draw that object.
- The first team finished shows their picture to the announcer who decides if the picture is accurate. If the picture is correct, that team earns one point. The team with the most points wins!

Be creative and make your own description cards to add to the pile.

1. A regular polygon with six sides.	2. A 3-sided figure with one right angle.	3. A quadrilateral with four right angles.	4. A 6-sided figure with different side lengths.	5. A 3-sided figure with equal angle measures.
6. A figure with four sides.	7. A figure with four congruent sides and four right angles.	8. A figure with 8 congruent sides.	9. A figure with 2 pairs of parallel sides and opposite congruent angles.	10. A 4-sided figure with only one pair of parallel sides.
11. A parallelogram with four congruent sides.	12. A rectangle with four congruent sides.	13. A parallelogram that is not a rectangle or a square but has diagonals that are perpendicular.	14. A 5-sided figure with congruent angles and sides.	15. A 3-sided figure with two congruent sides.

Answers: 1. Regular hexagon **2.** Right triangle **3.** Rectangle **4.** Irregular hexagon **5.** Equilateral triangle **6.** Quadrilateral **7.** Square **8.** Regular Octagon **9.** Parallelogram **10.** Trapezoid **11.** Rhombus **12.** Square **13.** Rhombus **14.** Regular pentagon **15.** Isosceles triangle

CHAPTER 8 — Family Letter
Section C

What We Are Learning

Polygon Relationships

Vocabulary
These are the math words we are learning:

line of reflection a line over which a figure flips in order to create a mirror image of the original figure

line symmetry occurs when a figure can be reflected or folded so the two parts of the figure match or are congruent

line of symmetry the line over which a figure is folded or flipped to create line symmetry

reflection occurs when a figure is flipped over a line of reflection

rotation the movement of a figure around a point

transformation a change in the size or position of a figure

translation the movement of a figure along a straight line

Dear Family,

Now that your child has a solid understanding of basic geometric concepts, he or she will begin to explore the relationships that can be applied to polygons. One such application is the concept of congruence.

Your child studied congruent angles in the previous sections. Now this concept will be applied to figures. Figures are congruent when they have the same size and the same shape. Your child will learn to identify if a pair of figures are congruent.

Decide whether the two figures in each pair are congruent. If not, explain.

A.

The figures are both trapezoids. They are neither the same shape nor the same size. The figures are not congruent.

B.

 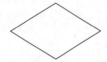

These figures have the same shape and size. The figures are congruent.

Another geometric concept that can be applied to polygons is **symmetry.** If a figure is folded or reflected and the two parts of the figure match, then the figure is symmetrical. The line of reflection is also called the **line of symmetry.**

Determine whether each dashed line appears to be a line of symmetry.

A.

B.

The two parts of the figure are congruent, but do not match exactly when reflected over the line. The line does not appear to be a line of symmetry.

The two parts of the arrow appear to match exactly when reflected over the line. The line appears to be a line of symmetry.

Family Letter
Section C, continued

When a figure is moved along a plane and does not change its shape or size, it is called a **transformation.** There are three types of transformations: translations, reflections, and rotations.

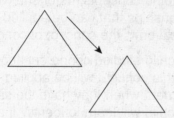

A **translation** moves the figure along a straight line. Only the location changes when a figure is translated.

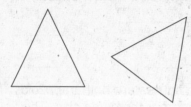

In a **rotation,** the figure rotates around a point. Both the location and the position change during a rotation.

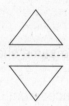

With a **reflection,** the figure is flipped over a line, called the **line of reflection,** which creates a mirror image of the figure. The location and the position of the figure change during a reflection.

Review these geometric applications often with your child. Encourage your child to find transformations, symmetry, and tessellations in everyday situations. Being able to connect these ideas and concepts with real life examples will be helpful to your child.

Sincerely,

Name _____ Date _____ Class _____

CHAPTER 8 Family Letter
Polygon Relationships

Decide whether the figures in each pair are congruent. If not, explain.

1.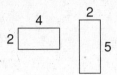

2.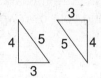

_____ _____

Tell whether each is a translation, rotation, or reflection.

3.

4.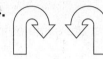

_____ _____

Determine whether each dashed line appears to be a line of symmetry.

5.

6.

_____ _____

Name the coordinate of each vertex when the transformation is made.

7.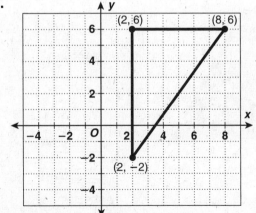

reflected over the y-axis

8.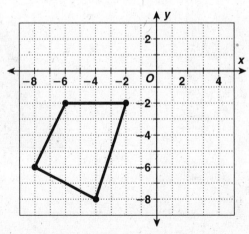

translated 4 to the right and 2 up

Answers: 1. The figures are not congruent. Sides are not the same size. 2. The figures are congruent. 3. Translation 4. Reflection 5. No. 6. Yes 7. (−2, −2) (−2, 6) (−8, 6) 8. (−2, 0) (2, 0) (−4, −4) (0, −6)

71 Holt Mathematics

Name _____ Date _____ Class _____

CHAPTER 8

Family Fun
Art Fun

Geometric concepts and applications of symmetry, reflection, and tessellations are often used in drawings and paintings. Become a mathematical artist by creating your own work of art!

Materials

- Magazines and/or pictures
- Glue and scissors
- Large sheets of construction paper

Directions

- Look for real life images that are symmetrical in your magazines or pictures.
- Cut these images out and put them off to one side.
- Categorize the pictures as to whether they have a horizontal or vertical line of symmetry.
- Create a collage with all of the images you found that have the same type of line of symmetry. Glue these images onto the construction paper.

Share your masterpiece with your family.

For example, these images all have a vertical line of symmetry. Can you identify each one?

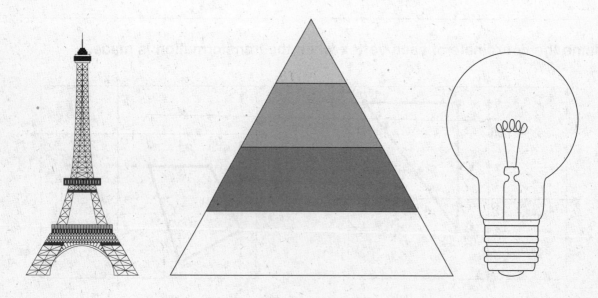

Family Letter
Chapter 9, Section A

What We Are Learning

Customary and Metric Measurement

Vocabulary
These are the math words we are learning:

customary system a system of measurement primarily used in the United States

metric system a system of measurement used all over the world whose units are related by the decimal system

Dear Family,

Your child will be learning about the customary system of measurement. There are different units of measure for length, weight, and capacity. Length can be measured by the inch, the foot, the yard, or the mile. Weight is measured by the ounce, the pound, and the ton. Capacity is measured by the fluid ounce, the cup, the pint, the quart, and the gallon.

While the customary system is used primarily in the United States, other parts of the world use the metric system. This system has different units of measurement that are comparable but not equivalent to the units in the customary system. The metric units of length are millimeters, centimeters, meters, and kilometers. The metric units used to measure mass are milligrams, grams, and kilograms. Capacity is measured in the metric system by milliliters, and liters.

Your child will practice choosing the appropriate units or measurement tools necessary to measure different objects. For example, the length of a certain road would be measured in miles or kilometers, not inches or centimeters. The capacity of a drinking glass would be measured in milliliters or fluid ounces, not liters or gallons. Practice with your child at home by choosing household objects and asking in what units they should be measured.

Your child will also learn how to convert between units in each measuring system. The table below shows the conversions for the customary system.

Length	12 in. = 1 foot; 3 ft = 1 yard; 5,280 ft = 1 mile
Weight	16 oz = 1 lb; 2,000 lb = 1 T
Capacity	8 fl oz = 1 cup; 2 c = 1 pint; 2 pt = 1 quart; 2 qt = 1 gallon

Family Letter

Section A, continued

Notice that some measurement conversions can be made with more than one unit. For example 5,280 ft = 1 mi and since 3 ft = 1 yd, 1,760 yd = 1 mi. This demonstrates how we are able to change, or convert, between different units. To convert a measurement from larger units to smaller units, multiply the measurement with larger units by the number of smaller units it takes to make one of the larger unit; to find feet in inches, multiply by 12. To convert from smaller to larger units, divide the quantity of the measurement in smaller units by the number of smaller units it takes to make one of the larger unit; to convert ounces to pounds, divide by 16.

The metric system works in much the same way. The conversions are below.

Length	10 millimeters = 1 centimeter; 100 cm = 1 meter; 1000 m = 1 kilometer
Mass	1000 milligrams = 1 gram; 1000 g = 1 kilogram
Capacity	1000 milliliters = 1 liter; 1000 l = 1 kiloliter

You may notice that the conversions for the metric system are similar. You can use the same methods of multiplying and dividing from the customary system to convert in the metric system. However, since all these measurements are based on powers of ten, there is a shortcut to conversion in the metric system. To convert to smaller units, move the decimal point to the right based on the number of zeros in the conversion factor. To convert to larger units, move the decimal point to the left. For example:

4.850 kg = 4,850 g   576 cm = 5.76 m

1000 g in 1 kg = 3 zeros   100 cm in 1 m
move decimal point 3 places   move decimal point 2 places
right   left

Your child will also learn about units of time and temperature. There is only one system for measuring time. It is measured in seconds, minutes, and hours. 60 of the smaller units make up 1 of the larger units. There are two systems for measuring temperature: Fahrenheit and Celsius. Both are measured in degrees. Water freezes at 0° C or 32° F and boils at 100° C or 212° F.

Help your child become familiar with these units and conversions. Point out measurements such as capacity of a container or distance on a map and ask how many of another unit that measurement would equal.

Sincerely,

Name _____ Date _____ Class _____

Family Letter
Chapter 9 — Customary and Metric Measurement

Name the appropriate units to measure each.

Customary
1. the length of a football field _____
2. the capacity of a can of juice _____
3. the weight of an elephant _____

Metric
4. the mass of a pair of scissors _____
5. the length of an eyelash _____
6. the capacity of a bucket _____

Convert.

7. 4 mi to ft
8. 675 l to ml
9. 48 oz to lb

10. 3.762 kg to g
11. 64 pt to gal
12. 4,391 mm to m

13. 2.47 T to lb
14. 5.681 l to ml
15. 60 in. to ft

16. 9.753 m to cm
17. 6 gal to c
18. 4.8 mg to g

19. 1.5 hours to minutes
20. 45 minutes to hours

Find the temperature.

21.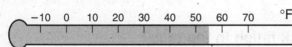

22.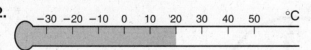

Answers: 1. yards 2. fluid ounces 3. tons 4. grams 5. millimeters 6. kiloliters 7. 21,120 ft 8. 675,000 ml 9. 3 lb 10. 3,762 g 11. 16 gal 12. 4.391 m 13. 4,940 lb 14. 5,681 ml 15. 5 ft 16. 975.3 cm 17. 48 c 18. 0.0048 g 19. 90 min 20. 0.75 hr 21. 55° F 22. 20° C

Name _____ Date _____ Class _____

CHAPTER 9
Family Fun
Measurement Jumble

Fill in the answers. Then complete the riddle below by using the selected letter from each answer to fill in the blanks.

1. The appropriate metric unit to measure the capacity of a car's gas tank is the [ ] ___ ___ ___ .

2. One minute equals 60 ___ ___ [ ] ___ ___ ___ ___ .

3. Water freezes at 32° ___ ___ ___ ___ ___ [ ] ___ ___ ___ .

4. One cup equals 8 ___ ___ ___ ___ ___   ___ ___ [ ] ___ .

5. Move the decimal to the left to convert grams to ___ ___ ___ ___ ___ ___ [ ] ___ .

6. Temperature is measured in ___ ___ ___ ___ ___ ___ [ ] .

7. 10 millimeters equals 1 [ ] ___ ___ ___ ___ ___ ___ ___ .

8. The appropriate customary unit to measure the length of the classroom is the ___ ___ ___ [ ] .

9. The appropriate metric unit to measure the width of a pencil is the ___ ___ ___ ___ ___ [ ] ___ ___ .

10. 2,000 pounds equals 1 ___ [ ] ___ .

11. One gallon equals 8 [ ] ___ ___ ___ ___ .

Where did the clock finish in the race?

__ __ __ __ __ __ __ __ __ __ __
6 4 2 10 3 8 11 1 5 7 9

Answers: 1. liter 2. seconds 3. Fahrenheit 4. fluid ounces 5. kilograms 6. degrees 7. centimeter 8. yard 9. millimeter 10. ton 11. pints Riddle: second place

CHAPTER 9 Family Letter
Section B

What We Are Learning

Measurement and Geometric Figures

Vocabulary
These are the math words we are learning:

center the given point in a circle from which all points in a plane are the same distance

circle the set of all points in a plane that are the same distance from a given point, called the center

circumference the distance around the circle

diameter a line segment that passes through the center of a circle with endpoints on the circle

perimeter the distance around a polygon

pi the ratio of the circumference of a circle to the length of its diameter, $\frac{C}{d}$; represented by the Greek letter, π

radius (radii) a line segment with one endpoint at the center of the circle and the other endpoint on the circle

Dear Family,

Your child will learn about angle measures within polygons. The measures of the angles inside a triangle will always add to 180°. The angle measures in any quadrilateral will always equal 360°. Therefore, if you know all but one of the angle measures you can always find the last by using an algebraic expression. For example:

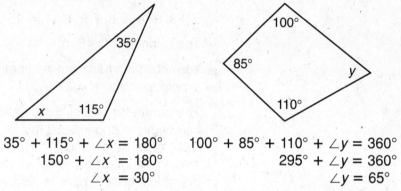

$35° + 115° + \angle x = 180°$ $100° + 85° + 110° + \angle y = 360°$
$150° + \angle x = 180°$ $295° + \angle y = 360°$
$\angle x = 30°$ $\angle y = 65°$

There are some special rules for regular quadrilaterals. Opposite pairs of angles in a regular quadrilateral are equal. Any two adjacent angles will total 180°. All angles in squares and rectangles are 90°. Therefore, by knowing only one angle of a regular quadrilateral, you can find all the angles.

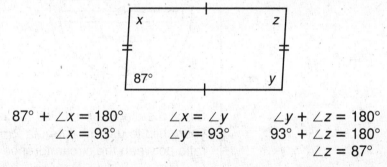

$87° + \angle x = 180°$ $\angle x = \angle y$ $\angle y + \angle z = 180°$
$\angle x = 93°$ $\angle y = 93°$ $93° + \angle z = 180°$
      $\angle z = 87°$

You can also tell that $\angle z = 87°$ because it is congruent with its opposite angle.

Your child will be learning to find the perimeter and area of different figures. The **perimeter** of a figure is the distance around the polygon. To find the perimeter your child will find the sum of the lengths of all sides of a figure.

Family Letter
Section B, continued

Find the perimeter of the figure.

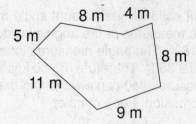

$5 + 8 + 4 + 8 + 9 + 11 = 45$

The perimeter is 45 m.

Your child may find the perimeter of a rectangle by using a simple formula: $P = 2\ell + 2w$.

The perimeter of a circle is called the circumference. Your child can begin to understand this concept by wrapping a flexible tape measure around a cylinder. The length the tape measures around the object is equal to the circumference of the cylindrical base. There is also a formula your child can use to find the circumference.

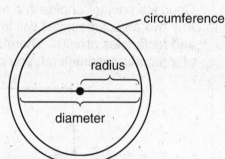

Circumference = πd

Since the diameter is twice the length of the radius, it may help your child to use the formula $2\pi r$. The pi symbol stands for the ratio between the circumference and the diameter and can be represented by the decimal 3.14 or the fraction $\frac{22}{7}$. In other words, the circumference is a little more than three times the diameter.

You can help your child study these concepts by providing objects for her or him to measure and providing "missing angle" problems for her or him to solve.

Sincerely,

Name _____ Date _____ Class _____

CHAPTER 9 Family Letter
Measurement and Geometric Figures

Find the measure of ∠x.

1.

_____

2.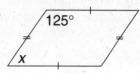

_____

3.

_____

4.

_____

Find the perimeter of each figure.

5.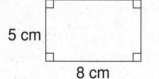

_____

6.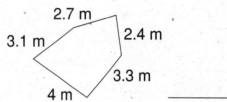

_____

Find each unknown measure.

7. The length of a rectangle is 11 meters. What is the perimeter of the rectangle if the width is 4 meters less than the length?

_____

8. An equilateral triangle has a perimeter of 18 cm. What is the length of each side?

_____

Find the circumference.

9.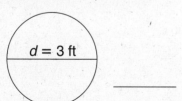

_____

10.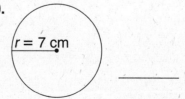

_____

Answers: 1. 65° 2. 55° 3. 105° 4. 60° 5. 26 cm 6. 15.5 m 7. 36 m 8. 6 cm 9. 9.42 ft 10. 43.96 cm

Name _____ Date _____ Class _____

CHAPTER 9
Family Fun
Perimeter and Circumference Match Up

Find the perimeter of each polygon and the circumference of each circle (round to the nearest tenth). All measurements are in centimeters. Then draw lines to match the circles and polygons whose perimeter and circumference most closely match.

1.

A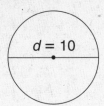

2.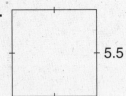

B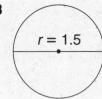

3.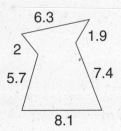

C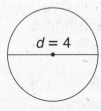

4.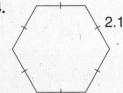

D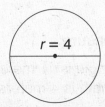

5.

E

Answers: 1. D 2. E 3. A 4. C 5. B

CHAPTER 10 Family Letter
Section A

What We Are Learning

Area

Vocabulary

These are the math words we are learning:

area the number of square units needed to cover a given surface

Dear Family,

Your child will learn the area formulas for rectangles, triangles, and parallelograms. The **area** of a figure is defined as the number of square units needed to cover a given surface.

Figure	Formula	Words
parallelogram with height and base labeled	$A = bh$	Area equals base times height.
triangle with height and base labeled	$A = \frac{1}{2}bh$	Area equals one-half the base times the height.
trapezoid with b_1, b_2, and h labeled	$A = \frac{(b_1 + b_2)h}{2}$	Area equals the sum of the lengths of the bases times the height divided by two.

Your child will be able to use these formulas to find the areas of composite figures. For example, the composite figure below can be divided into different groupings of recognizable polygons. You can find the area of each and combine the areas to find the area of the composite figure.

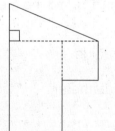

Copyright © by Holt, Rinehart and Winston.
All rights reserved.

Holt Mathematics

Chapter 10 Family Letter
Section A, continued

Your child will learn about the relationship between perimeter and area. Some figures can have the same perimeter while having different areas. For example:

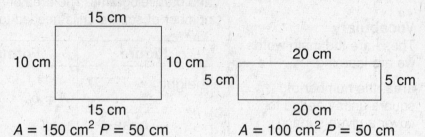

$A = 150 \text{ cm}^2 \; P = 50 \text{ cm}$ $A = 100 \text{ cm}^2 \; P = 50 \text{ cm}$

Your child may be given a set perimeter and asked to find the largest area that can be contained within it.

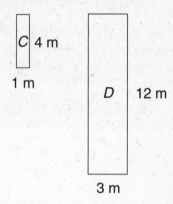

Another relationship between the area and perimeter of a figure is shown when enlarging a figure. If the dimensions of a figure are all multiplied by the same number, the area increases to equal the area of the smaller figure multiplied by the square of the dimension multiplier.

For C: For D:
$A = 4 \text{ m}^2$ $A = 4 \times 3^2 = 4 \times 9 = 36$
     $A = 3 \times 12 = 36$

Your child will also find the area of a circle. Your child will need to use this formula to find the area of a circle.

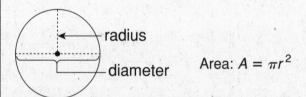

Area: $A = \pi r^2$

The ratio between the circumference of a circle and the diameter is called *pi*, represented by the symbol π. Your child may use the decimal approximation 3.14 or the fraction $\frac{22}{7}$ as a numeric substitution for *pi*.

Have your child explain the difference between the area, and perimeter or circumference of a figure. He or she will continue to use these formulas throughout most math programs.

Sincerely,

Name _____ Date _____ Class _____

CHAPTER 10 Family Letter
Area

Find the area of each polygon.

1.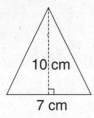
2.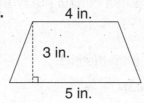
3.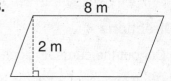

4.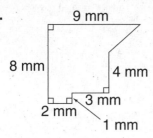
5.

6.
7.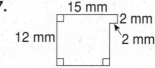

8. Brenda has a 3 inch × 5 inch photo of a sunset. If she enlarges this photo so it will fit in a 12 inch × 20 inch frame, how will its area change?

Find the area and circumference of each circle. Use $\frac{22}{7}$ for pi.

9.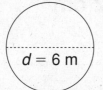
10.

Answers: 1. 35 cm² 2. 13.5 in² 3. 16 m² 4. 43 mm² 5. 28 ft² 6. 15.05 m² 7. 160 mm² 8. It will increase 16 times. 9. 28.29 m²; 18.86 m 10. 50.29 cm²; 25.14 cm

Name _____ Date _____ Class _____

CHAPTER 10

Family Fun
Dimension Creation

Shapes may cover the same area, but have different perimeters. How many different figures can you create with a certain area? Be creative!

Directions

- Cut out the cards. Pick an area card from the pile.
- Draw and label 3 figures with that area.
- Calculate the perimeters of each. You earn 1 point for each perimeter that is correctly calculated.
- You earn 1 point for each quadrilateral, 3 points for each triangle, and 5 points for each circle that covers that particular area.
- The person with the most points is the winner.

Perimeter =	Perimeter =	Perimeter =

24 m²	18 m²	36 m²
100 cm²	42 cm²	12 cm²
56 in²	8 in²	25 in²

Copyright © by Holt, Rinehart and Winston.
All rights reserved.

Holt Mathematics

Family Letter
Chapter 10, Section B

What We Are Learning

Volume and Surface Area

Vocabulary
These are the math words we are learning:

base a face of a three-dimensional figure that usually determines the name of the figure

cone a solid figure with a circular base and a curved surface that comes to a point

cylinder a solid figure with two parallel, congruent, circular bases connected by a curved lateral surface

edges the intersection of two faces of a polyhedron

face the flat sides of a polyhedron

net an arrangement of two-dimensional figures that folds to form a polyhedron

polyhedron a solid figure in which all the surfaces or faces are polygons

prism a polyhedron with two congruent, parallel bases, and parallelogram faces

pyramid a polygon-shaped base with triangular faces that come to a point named for the shape of its base

surface area the sum of the areas of the faces of a solid figure

Dear Family,

Your child will continue learning about geometric shapes. In this section, your child will be introduced to solid figures. Solid figures are three-dimensional, which means they have depth, as well as length and height.

Your child will learn to study three-dimensional figures from different views. When presented with three different views of objects, front, side, and top, your child will be able to visualize and draw the three-dimensional object.

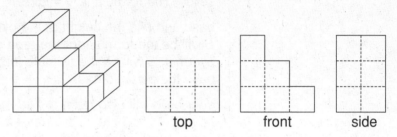

A special type of solid figure is a **polyhedron.** The sides, or **faces,** of a polyhedron are polygons. The **edges** are formed by two faces, and a **vertex** is formed by three or more edges.

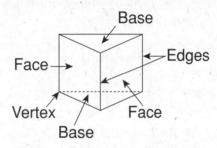

A **prism** is a special polyhedron that has two congruent, parallel bases, and parallelogram faces. Prisms are named for the shape of their bases. A **pyramid** has a polygon shaped base, but the other faces are triangles. It is also named for its base.

   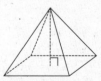

Rectangular Prism   Pentagonal Prism   Triangular Prism   Rectangular Pyramid

Cylinders and **cones** are NOT polyhedra because not every surface is a polygon.

Cylinder   Cone

Family Letter
Section B, continued

vertex the point at which three or more edges of a figure meet

volume the number of cubic units needed to fill a space

Your child will use these volume formulas to find the volume of each solid figure.

Rectangular Prism	Triangular Prism	Cylinder
$V = \ell w h$	$V = Bh$	$V = \pi r^2 h$

Cones and pyramids have the same formula for finding their volumes, which is slightly different from the formulas for other solids. The formula is $V = \frac{1}{3}(Bh)$, where B is the area of the base and h is the height of the pyramid. In the case of a cone, B will be the formula for the area of a circle, whereas in a pyramid, B will be the formula for the area of whatever type of base the pyramid has.

Your child will also learn to find the **surface area** of many solid figures. To find the surface area of a prism, your child can use a net. A **net** is the pattern made when the solid is laid out flat showing each face of the solid. To find the surface area, add the areas of each face.

Your child will use these formulas to find the surface area of a pyramid and a cylinder.

Pyramid	Cylinder
$S = s^2 + 4\left(\frac{1}{2}bh\right)$	$S = h(2\pi r) + 2(\pi r^2)$

Your child will use these formulas in most math classes. These formulas are also used in many real-life situations.

Sincerely,

Name _____ Date _____ Class _____

CHAPTER 10 Family Letter
Volume and Surface Area

1. Draw top, side, and front views of the solid.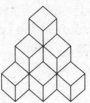

Identify the number of faces, edges, and vertices in each solid figure.

2.  _____

3. 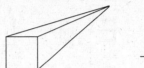 _____

Tell whether each figure is a polyhedron and name the solid.

4. 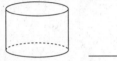 _____

5. 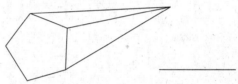 _____

Find the surface area of each figure.

6. 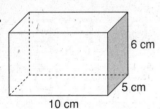 _____

7. 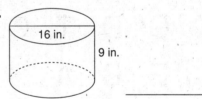 _____

Find the volume of each solid.

8. _____

9. 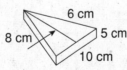 _____

10. 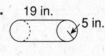 _____

11. 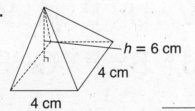 _____

12. 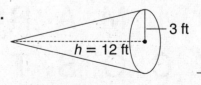 _____

Answers: 1. [top, side, front views] 2. 5 faces 9 edges 6 vertices 3. 5 faces 8 edges 5 vertices 4. no; cylinder 5. yes; pentagonal pyramid 6. 280 cm² 7. 854.1 in² 8. 336 m³ 9. 200 cm³ 10. 1,491.5 in³ 11. 32 cm³ 12. 113.04 ft³

Copyright © by Holt, Rinehart and Winston.
All rights reserved.

Holt Mathematics

Name _____ Date _____ Class _____

Chapter 10 Family Fun
Word Find

Directions
Unscramble the vocabulary words in the box.
Find those words hidden in the puzzle.

| enco | derlcyin | cefa | smipr | xeevtr | eegd |
| oornehylpd | yrmdiap | tne | frsucea aare | mlveou | seba |

```
S O E I I S B C N M P R
F U E M K P Y W S Y O C
U P R F U L J I D N L C
D U T F I L R R F X Y M
S O Z N A P O S A E H Y
K C D T O C S V C T E Y
C E B A S E E H E R D W
R O V P D L V A P E R O
C T N G T E N R R V O F
W I E E M K M I F E N D
D I M A R Y P V U L A W
T C O S T O D J B F Q R
```

Answer: cone, cylinder, face, prism, vertex, edge, polyhedron, pyramid, net, surface area, volume, base

Chapter 11 Family Letter
Section A

What We Are Learning

Understanding Integers

Vocabulary
These are the math words we are learning:

absolute value the distance from 0 on a number line. The symbol for absolute value is | |

coordinates the numbers in an ordered pair that locate a point on a coordinate graph

coordinate plane formed by two number lines in a plane that intersect at right angles at zero on each number line

integers the set of all whole numbers and their opposites

negative number integer less than zero

opposites two numbers that are an equal distance from zero on a number line

origin the point where the x-axis and y-axis intersect on the coordinate plane

positive number integer greater than zero

quadrants the four areas created by the axes on a coordinate plane

x-axis the horizontal axis on the coordinate plane

x-coordinate the first value in an ordered pair

y-axis the vertical axis on the coordinate plane

Dear Family,

Your child will begin the study of an important set of numbers called **integers**. Integers are the set of numbers that include both positive and negative whole numbers. Your child will learn to identify and graph these numbers on a number line.

Name a positive or negative number to represent each situation.

A. a gain of 2 points in the stock market

Positive numbers can represent gains or increases.
+2

B. a 15 yard penalty in football

Negative numbers can represent losses or decreases.
−15

To help your child understand integers, it is helpful to graph an integer and its opposite on a number line. **Opposites** are two numbers that are an equal distance from zero on a number line.

Graph the integer 6 and its opposite on a number line.

+6 is the same distance from 0 as −6.

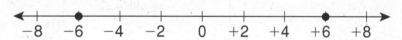

Absolute value represents the distance an integer is from zero. The absolute value of a number is always positive because it reflects a distance, which is always a positive value. Therefore, opposites always have the same absolute value.

Use the number line to find the absolute value of |−3|.

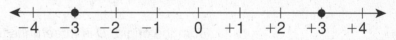

−3 is 3 units from 0, so |−3| = 3.

CHAPTER 11 Family Letter
Section A, continued

y-coordinate the second value in an ordered pair

Your child will also learn to compare and order integers. It is important to remember that negative numbers are **ALWAYS** less than positive numbers.

Use the number line to compare each pair of integers. Write < or >.

−5 ☐ 3

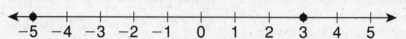

−5 is to the left of 3 on the number line.

−5 < 3

Once your child has a basic understanding of integers, he or she will be introduced to the **coordinate plane.** He or she will plot points or identify points in all four **quadrants** of the coordinate plane. For example:

Give the coordinates of point D.

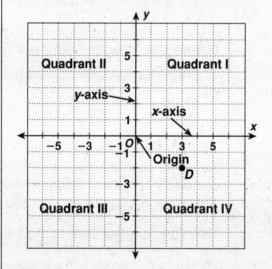

From the origin, D is three units to the right and 2 units down.

(3, −2)

This is just the beginning of your child's involvement with integers. Understanding the concept of positive and negative values is imperative as your child continues in mathematics.

Sincerely,

Name _____ Date _____ Class _____

CHAPTER 11 Family Letter
Understanding Integers

Name a positive or negative number to represent each situation.

1. a loss of 15 yards in football _____
2. an increase of 52 points _____

Graph each integer and its opposite on a number line.

3. 10

<----|---|---|---|---|---|---|---|---|---|---|---->
−10−8−6−4−2 0 2 4 6 8 10

4. −8

<----|---|---|---|---|---|---|---|---|---|---|---->
−10−8−6−4−2 0 2 4 6 8 10

Use the number line to find the absolute value of each integer.

<----|---|---|---|---|---|---|---|---|---|---|---|---|---|---|---|---|---|---|---|---|---->
−10−9−8−7−6−5−4−3−2−1 0 1 2 3 4 5 6 7 8 9 10

5. |−4| 6. |6| 7. |−9|

_____ _____ _____

Use the number line to compare each pair of integers. Write < or >.

<----|---|---|---|---|---|---|---|---|---|---|---|---->
−6 −5 −4 −3 −2 −1 0 1 2 3 4 5 6

8. 0 ___ −5 9. −3 ___ 5 10. −2 ___ −4

Order the integers in each set from least to greatest.

11. −6, 2, −4 12. 0, −5, −7 13. −9, 8, 1, −1

_____ _____ _____

Use the coordinate plane to answer Exercises 14 − 19.

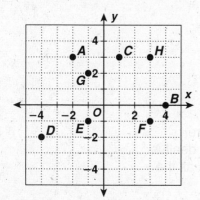

Name the quadrant where each point is located.

14. D 15. H 16. A

_____ _____ _____

Give the coordinates of each point.

17. C 18. B 19. F

_____ _____ _____

Answers: 1. −15 2. +52 3. <number line> 4. <number line> 5. 4 6. 6 7. 9 8. > 9. < 10. > 11. −6, −4, 2 12. −7, −5, 0 13. −9, −1, 1, 8 14. III 15. I 16. II 17. (1, 3) 18. (4, 0) 19. (3, −1)

91   Holt Mathematics

Name _____ Date _____ Class _____

Family Fun
Integer Game

Objective
The objective is to collect all of the cards in the deck.

Materials
Deck of cards

Directions
Deal out the entire deck to every person playing in the game. Make sure the cards are face down.

Every round is played the same way: each player turns over the top card in his or her hand. The player who has the card with the greatest value takes all the cards played that round.
However…

- Each **black** card holds a positive value with the aces being the least value and the kings being the greatest value.

- Each **red** card holds a negative value with the aces having the greatest value and the kings having the least value.

For example: If a red 7 and a black 2 are turned over, the player with the black 2 will win that round because $2 > -7$.

If there is a tie, or a round where two cards with the same value are turned up, each player deals out 3 cards face down. The last card dealt is turned over. The player with the card with the greatest value takes all the cards that round.

The winner is the player who ends up with all the cards or who has the greatest number of cards at the end of the game.

Copyright © by Holt, Rinehart and Winston.
All rights reserved.

Holt Mathematics

CHAPTER 11 Family Letter
Section B

What We Are Learning

Integer Operations

Dear Family,

In this section, your child will learn to perform the basic operations of addition, subtraction, multiplication, and division on integers.

Your child will use a number line to help add and subtract integers.

Integer Addition	Integer Subtraction
Move **right** on the number line to add a positive integer.	Move **left** on the number line to subtract a positive integer.
Move **left** on the number line to add a negative integer.	Move **right** on the number line to subtract a negative integer.

Find the sum.

$8 + (-2)$

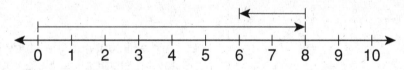

$8 + (-2) = 6$

Find the difference.

$-8 - (-3)$

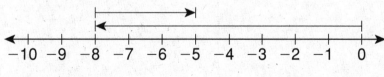

$-8 - (-3) = -5$

Holt Mathematics

Family Letter
Chapter 11 Section B, continued

The rules for multiplying and dividing integers are easier to use and to remember. The sign and/or signs of the integers will dictate which sign the product or quotient will have.

- If the signs are the same, the product or quotient will be **POSITIVE**.
- If the signs are different, the product or quotient will be **NEGATIVE**.

Find each product or quotient.

A. $6 \cdot (-3)$ Multiply.

$6 \cdot (-3) = -18$ The signs are different, so the answer is negative.

B. $-56 \div (-8)$ Divide.

$-56 \div (-8) = 7$ The signs are the same, so the answer is positive.

Your child will use the basic operations to solve simple equations involving integers.

Solve.

$-7a = 35$

$\dfrac{-7a}{-7} = \dfrac{35}{-7}$ a is multiplied by -7. Divide both sides by -7.

$a = -5$

Your child will continue the study of integers throughout all aspects of mathematics. Help your child master these important skills by reviewing the concepts taught throughout this chapter.

Sincerely,

Name _____ Date _____ Class _____

CHAPTER 11 Family Letter
Integer Operations

Find each sum.

1. $-8 + 3$

2. $6 + (-6)$

3. $4 + 7$

Evaluate $-2 + x$ for each value of x.

4. 5

5. -2

6. -7

Find each difference.

7. $1 - 9$

8. $-5 - 6$

9. $8 - (-4)$

Evaluate $a - (-5)$ for each value of a.

10. -3

11. 2

12. -7

Find each product.

13. $0 \cdot 12$

14. $-6 \cdot 3$

15. $-4 \cdot -7$

Evaluate $-6m$ for each value of m.

16. -1

17. 5

18. -6

Divide.

19. $\dfrac{36}{-4}$

20. $-8 \div -2$

21. $-56 \div 7$

Solve each equation.

22. $x - 13 = -5$

23. $15 + a = 4$

24. $\dfrac{w}{-3} = 6$

Answers: 1. -5 2. 0 3. 11 4. 3 5. -4 6. -9 7. -8 8. -11 9. 12 10. 2 11. 7 12. -2 13. 0 14. -18 15. 28 16. 6 17. -30 18. 36 19. -9 20. 4 21. -8 22. $x = 8$ 23. $a = -11$ 24. $w = -18$

Name _____ Date _____ Class _____

CHAPTER 11 Family Fun
Integer Fill In

Directions
- Try to fill in the missing numbers.
- The missing numbers are integers between −100 and 100.
- The numbers in each row add up to the total in the right column.
- The numbers in each column add up to the total along the bottom row.
- The diagonal lines add up to the total in the upper right and lower right cells.

									−158
	15	−74	−55	54		−54		−47	−52	−293
−28	97	−29		83	−73	−30	62	−4		24
				14	35				93	−43
		−62	−96		11		−75	−60		−299
3	−15	42	−84	−12	89	−29		−10	−41	−92
8	−6	−3		81		91	93	34		331
	−23		−36	−6	99		−94		13	−168
92	−15		3	−29	86		66	76	86	411
−49	−76		−16	29	−73	23	22	−28	−2	−110
−42	−99	−86	31	75	77	−48	−59		−50	−230
−83	−43	−241	−279	231	84	−129	−131	−35	157	−181

Family Letter
Chapter 11 Section C

What We Are Learning

Functions and Equations

Vocabulary
These are the math words we are learning:

function a rule that relates two quantities so that each input value corresponds exactly to one output value

input a value substituted into a function to get an output

linear equation an equation whose graph is a straight line

output a result of a function after the input value is placed in the function

Dear Family,

Your child will be learning about **functions** and how to write equations for a function. A function is simply just a rule for a pattern. It relates two quantities so that each **input** value corresponds *exactly* to only *one* **output** value.

You can use a function table to show some of the values for the function, as shown in this example.

Write an equation for the function that gives the values in the table. Use the equation to find the value of y for the indicated value of x.

x	5	6	7	8	9	10
y	13	15	17	19	21	■

y is 2 times $x + 3$.	Compare x and y to find a pattern.
$y = 2x + 3$	Use the pattern to write an equation.
$y = 2(10) + 3$	Substitute 10 for x.
$y = 20 + 3$	Multiply.
$y = 23$	Use your function rule to find y when $x = 10$.

When x is 10, y is 23.

Another important part of this section is learning how to translate words into math expressions.

Write the equation for the function. Tell what each variable you use represents.

The height of a triangle is 6 less than the length of the base.

h = height of the triangle	Choose variables for the equation.
b = length of the base.	
$h = b - 6$	Write an equation.

Chapter 11 Family Letter
Section C, continued

As an introduction to functions, your child will learn about linear functions; a function represented by a straight line. She or he will learn to use ordered pairs and graphs to identify and represent various linear functions.

Use the given *x*-values to write solutions of the equation as ordered pairs. Then, graph the function described by the equation.

$y = 4x + 5$ for $x = 1, 2, 3, 4$

Make a function table by using the given values for x to find values for y.

x	4x + 5	y
1	4(1) + 5	9
2	4(2) + 5	13
3	4(3) + 5	17
4	4(4) + 5	21

Write these solutions as ordered pairs.
(x, y)
(1, 9)
(2, 13)
(3, 17)
(4, 21)

Graph the ordered pairs on a coordinate plane.

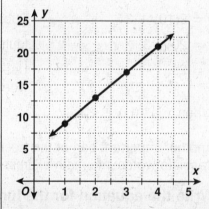

Draw a line through the points to represent all the values of x you could have chosen and the corresponding values of y.

Your child will also determine if an ordered pair is a solution of an equation by substituting the ordered pair into the equation to see if the values make the equation true.

While this is just an introduction to linear functions and equations, your child will continue to build upon the information learned in this section throughout future math courses.

Sincerely,

Name _____ Date _____ Class _____

CHAPTER 11 Family Letter
Functions and Equations

Write an equation for a function that gives the values in each table. Use the equation to find the value of y for the indicated x.

1.
x	2	3	4	5	6	10
y	3	5	7	9	11	■

_____

2.
x	2	3	5	7	12
y	8	9	11	13	■

_____

Write an equation for the function. Tell what each variable you use represents.

3. Mark works at the market. For each hour he works, Mark gets paid $6.50.

_____

4. Haley is five years older than Emma.

_____

Use the given x-values to write solutions of each equation as ordered pairs.

5. $y = 3x - 2$ for $x = 1, 2, 3, 4$

_____

6. $y = -6x + 4$ for $x = 1, 2, 3, 4$

_____

Determine whether each ordered pair is a solution of the given equation.

7. $(2, 5); y = 6x - 5$

_____

8. $(-4, 1); y = x + 5$

_____

Use the graph of the linear function to find the value of y for each given value of x.

9. $x = -3$ 10. $x = 0$

_____   _____

11. $x = 1$ 12. $x = 3$

_____   _____

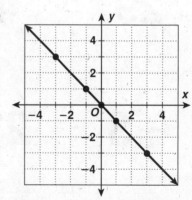

Answers: 1. $y = 2x - 1$; $y = 19$ 2. $y = x + 6$; $y = 18$ 3. $x =$ number of hours worked; $y =$ Mark's paycheck; $y = 6.50x$ 4. $x =$ Emma's age; $y =$ Haley's age; $y = x + 5$ 5. $(1, 1); (2, 4); (3, 7); (4, 10)$ 6. $(1, -2); (2, -8); (3, -14); (4, -20)$ 7. No; $5 \ne 6(2) - 5; 5 \ne 7$ 8. Yes; $1 = -4 + 5; 1 = 1$ 9. $y = 3$ 10. $y = 0$ 11. $y = -1$ 12. $y = -3$

Name _____ Date _____ Class _____

CHAPTER 11
Family Letter
Function Fun

Directions

- Cut out the cards and place them face down in rows and columns.
- Take turns with a partner uncovering 2 cards at a time.
- The first card is the *x*-value and the second card is the *y*-value.
- Substitute these values into the equation $y = -6x + 5$. If the values are a solution to the equation, you have a match. If the values are not a solution to the equation, turn the cards over and let the other player have a turn.
- The first player to find 3 matches is the winner.

5	1	−1	−1
−2	17	2	−7
29	4	−19	3
−3	23	$\frac{1}{2}$	2
0	6	−4	−13

Copyright © by Holt, Rinehart and Winston.
All rights reserved.

Holt Mathematics

CHAPTER 12 Family Letter
Section A

What We Are Learning

Understanding Probability

Vocabulary
These are the math words we are learning:

complement a possible outcome that is not what you expect or hope to get

equally likely all outcomes have an equal chance of occurring

experiment any activity based on chance

experimental probability The ratio of the number of times an event occurs to the total number of times the experiment is performed

fair an experiment with equally likely outcomes

outcome the possible results that can occur in an experiment

probability the measure of how likely an event is to occur

sample space the set of all possible outcomes in an experiment

theoretical probability the ratio of the number of ways an event can occur to the number of possible outcomes

Dear Family,

In this section, your child will be finding how to determine the probability of an event. **Probability** is defined as the measure of how likely an event is to occur.

You can write a probability as a fraction, a decimal between 0 and 1, or a percent between 0% and 100%. If an event has a probability close to 1 or 100%, the likelihood of the event to occur is much greater than if that same event has a probability close to 0 or 0%.

The greater the probability, the more likely the event will occur.

Write *impossible, unlikely, as likely as not, likely,* or *certain* to describe each event.

A. You roll an 8 on a standard number cube.

impossible

B. There are equal amounts of red and blue marbles in a bag and you pick a red marble.

as likely as not

C. There are 7 days in one week.

certain

Next, your child will use a ratio to find the **experimental probability** of an event, the number of times an event occurs to the total number of times the experiment is performed. In an **experiment**, the **outcomes** are the many different results that may occur, while the **sample space** is a list of ALL possible outcomes.

Janie practiced her 3-point shot and recorded how many times she scored. Find the experimental probability that Janie will make a 3-point shot.

Outcome	Made the shot	Missed the shot
Frequency	|||| |||| |||| |||| |||| |||| |||| ||	|||| |||| |||

$P(\text{makes a shot}) \approx \dfrac{\text{number of times the event occurred}}{\text{total number of trials}} = \dfrac{37}{50}$

The experimental probability of Janie making a shot is $\dfrac{37}{50}$.

Family Letter

Chapter 12 — Section A, continued

Your child will also learn to find the **theoretical probability** of an event. Theoretical probability is used when all the outcomes have an equal, or **fair,** chance of occurring. You can use this ratio to calculate the theoretical probability of an event.

$$\text{probability} \approx \frac{\text{number of ways an event can occur}}{\text{total number of possible outcomes}}$$

What is the probability of a 4 being rolled on a fair number cube?

There are six possible outcomes when rolling a fair number cube: 1, 2, 3, 4, 5, and 6. All are equally likely to occur.

$$P(\text{a 4 is rolled}) = \frac{?}{6 \text{ possible outcomes}}$$

There is only 1 way for a 4 to occur.

$$P(\text{a 4 is rolled}) = \frac{1 \text{ way to roll a 4}}{6 \text{ possible outcomes}}$$

$$P(\text{a 4 is rolled}) = \frac{1 \text{ way to roll a 4}}{6 \text{ possible outcomes}} = \frac{1}{6}$$

Knowing how to calculate probabilities is an important skill. Probabilities are used in many different jobs and situations. Have your child identify the many different scenarios where probabilities are being used or have been used.

Connecting these concepts to real life events will help your child to see how math is used everyday.

Sincerely,

Name _____ Date _____ Class _____

CHAPTER 12 Family Letter
Understanding Probability

Write *impossible*, *unlikely*, *equally likely*, *likely*, or *certain* to describe each event.

1. You roll a seven on a standard number cube.

   _____

2. You correctly guess one winning number between 1 and 1,000.

   _____

3. August will come after July.

   _____

For each experiment, identify the outcome shown and the sample space.

4. _____

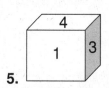

5. _____

Robert has a bag of marbles. He removes one marble, records the color, and places it back in the bag. He repeats this process several times and records the results in the table as shown.

yellow	|||| |||| ||
purple	|||| |||| ||||
red	|||| ||||	
green	|||| |||| |||

6. Find the probability that a marble selected from the bag is purple.

   _____

7. Find the probability that a marble selected from the bag will not be yellow.

   _____

Solve.

8. What is the probability of rolling a number that is a multiple of 2 on a fair number cube?

9. Suppose there is a 35% chance of rain today. What is the probability that it will not rain?

Answers: 1. impossible 2. unlikely 3. certain 4. outcome shown: lite gray; sample space: white, lite gray, dark gray, black 5. outcome shown: 4; sample space: 1, 2, 3, 4, 5, and 6. 6. $\frac{19}{25}$ 7. $\frac{25}{25}$ 8. $\frac{1}{2}$ 9. 65%

CHAPTER 12

Family Fun
What's Your Probability?

Materials
scissors
box/bag

Directions

- Cut the pictures out and put them in a box or bag.
- Select one piece of paper and record the outcome in the table. Place the paper back into the box or bag. Repeat this process 100 times.

1. What is the sample space of the experiment?
   _____

2. What is the experimental probability of pulling out a heart?
   _____

3. What is the theoretical probability of pulling out a heart?
   _____

4. How do the experimental and theoretical probabilities compare?
   _____

Answers: 1. sun, star, smiley face, lightning bolt 2. Possible answer: $\frac{1}{5}$ 3. $\frac{5}{5}$ 4. Possible answer: The probabilities should be about the same.

Chapter 12 Family Letter
Section B

What We Are Learning

Using Probability

Vocabulary
These are the math words we are learning:

compound event an event made of two or more single events

prediction a guess about something in the future

population the whole group being surveyed

sample part of the group being surveyed

Dear Family,

In this section, your child will relate the concepts of probability to application problems. Sometimes your child may need to make an organized list to find all the possible outcomes of an event. One way to create an organized list is to make a tree diagram, as shown in this example.

Christine eats one fruit and one vegetable everyday at lunch. Her fruit choices are apples, oranges, bananas, and grapes. Her vegetable choices are corn, green beans, and broccoli. How many different combinations of fruits and vegetables can she have for her lunch?

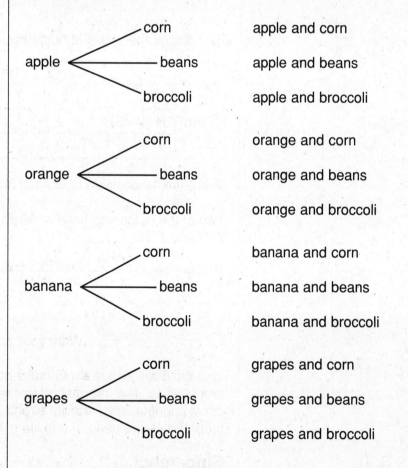

There are 12 different combinations of fruits and vegetables for Christine's lunch.

Holt Mathematics

Family Letter
Chapter 12, Section B, continued

Your child will also learn to find the probability of **compound events.** A compound event consists of two or more single events.

Carole flips a fair coin and spins the spinner.

Find the probability of the spinner showing an even number and the coin showing heads.

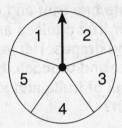

First find all the possible outcomes.

		Spinner				
		1	2	3	4	5
Coin	**H**	1, H	2, H	3, H	4, H	5, H
	T	1, T	2, T	3, T	4, T	5, T

There are 10 possible outcomes and all are equally likely.

Two of the outcomes have an even number and heads: 2, H and 4, H.

$$P(\text{even, heads}) = \frac{2 \text{ ways event can occur}}{10 \text{ possible outcomes}}$$

$$= \frac{2}{10}$$

$$= \frac{1}{5} \quad \text{Write your answer in simplest form.}$$

Your child will also learn to make predictions based on probabilities. The information learned in this chapter has a direct connection to real life events. Have your child share how probabilities are used in real life events.

Sincerely,

Name _____ Date _____ Class _____

CHAPTER 12 Family Letter
Using Probability

1. Central Middle School students wear uniforms every day. They have two choices of pants, navy or tan. They have three choices of shirts, white, red, or blue. How many different uniform combinations can the students wear?

   _____

2. Ms. Aylor's class is planning a class trip to the local museums and famous landmarks. The museums that they can visit are the Field Museum, History Museum, and the Aquarium. They also have the choice of visiting these famous buildings: the Capitol, the historical theater, and the governor's mansion. If they can visit only one museum and one landmark, how many different choices does the class have?

   _____

3. Three fair coins are flipped at the same time. What is the probability that two coins will show heads?

   _____

4. A bag contains six marbles: red, blue, green, black, white, and yellow. If you choose a marble and return it to the bag, what is the probability of picking green two times in a row?

   _____

5. You roll a fair number cube 50 times. How many times would you expect to roll a number that is a multiple of 2?

   _____

6. A survey of 150 people indicated that 45 of those surveyed eat five fruits or vegetables each day. Out of 1,250 people, predict how many people eat five fruits or vegetables each day.

   _____

Answers: 1. 6 2. 9 3. $\frac{3}{8}$ 4. $\frac{1}{36}$ 5. 25 times 6. 375 people

Name _____ Date _____ Class _____

CHAPTER 12 Family Fun
100% Fun

Materials
fair number cube
deck of cards
coin

Directions
- Cut out the cards and mix them up.
- Choose two cards and calculate the probability of each compound event.
- Add the probabilities after each round.
- The first person to reach 100% or have the greatest value after 5 rounds is the winner.

Challenge: Draw 3 or more cards and calculate the probabilities of the events occurring at the same time.

Coin Heads	Coin Not Tails	Coin Heads	Coin Tails	Coin Tails
Cards Drawing a red card	Cards Drawing a black ace	Cards Drawing a heart	Cards Drawing a black 3	Cards Drawing any face card
Fair number cube Rolling an even number	Fair number cube Rolling a number greater than 2	Fair number cube Rolling an odd number	Fair number cube Rolling an even number	Fair number cube Rolling any number but 3 or 4
Fair number cube Rolling a 7	Fair number cube Rolling a prime number	Fair number cube Rolling a 6	Fair number cube Rolling a 3	Fair number cube Rolling a 12

Holt Mathematics